MATHEMATICAL
MODELING

No. 7

Edited by
William F. Lucas, Claremont Graduate School
Maynard Thompson, Indiana University

Clark Jeffries

Code Recognition
and Set Selection
with Neural Networks

Birkhäuser
Boston • Basel • Berlin

Clark Jeffries
Department of Mathematical Sciences
Clemson University
Clemson, South Carolina 29634-1907
U.S.A.

Printed on acid-free paper.

© Birkhäuser Boston 1991

ISBN 0-8176-3585-8
ISBN 3-7643-3585-8

Camera-ready text prepared by the authors.
Printed and bound by Edwards Brothers, Inc., Ann Arbor, Michigan.
Printed in the U.S.A.

9 8 7 6 5 4 3 2 1

Code Recognition and Set Selection with Neural Networks

TABLE OF CONTENTS

PREFACE   vii

Chapter 0--The Neural Network Approach to Problem Solving

    0.1 Defining a Neural Network   1
    0.2 Neural Networks as Dynamical Systems   2
    0.3 Additive and High Order Models   5
    0.4 Examples   6
    0.5 The Link with Neuroscience   8

Chapter 1--Neural Networks as Dynamical Systems

    1.1 General Neural Network Models   9
    1.2 General Features of Neural Network Dynamics   15
    1.3 Set Selection Problems   16
    1.4 Infeasible Constant Trajectories   22
    1.5 Another Set Selection Problem   22
    1.6 Set Selection Neural Networks with Perturbations   24
    1.7 Learning   26
    Problems and Answers   28

Chapter 2--Hypergraphs and Neural Networks

    2.1 Multiproducts in Neural Network Models   33
    2.2 Paths, Cycles, and Volterra Multipliers   35
    2.3 The Cohen-Grossberg Function   38
    2.4 The Foundation Function Φ 40
    2.5 The Image Product Formulation of High Order Neural
        Networks   42
    Problems and Answers   48

Chapter 3--The Memory Model

    3.1 Dense Memory with High Order Neural Networks   53
    3.2 High Order Neural Network Models   55
    3.3 The Memory Model   59
    3.4 Dynamics of the Memory Model   62
    3.5 Modified Memory Models Using the Foundation Function 70
    3.6 Comparison of the Memory Model and the Hopfield Model 73
    Problems and Answers   77

Chapter 4--Code Recognition, Digital Communications, and General Recognition

4.1 Error Correction for Binary Codes 82
4.2 Additional Tests of the Memory Model as a Decoder 84
4.3 General Recognition 88
4.4 Scanning in Image Recognition 92
4.5 Commercial Neural Network Decoding 94
Problems and Answers 96

Chapter 5--Neural Networks as Dynamical Systems

5.1 A Two-Dimensional Limit Cycle 102
5.2 Wiring 107
5.3 Neural Networks with a Mixture of Limit Cycles and Constant Trajectories 111
Problems and Answers 114

Chapter 6--Solving Operations Research Problems with Neural Networks

6.1 Selecting Permutation Matrices with Neural Networks 118
6.2 Optimization in a Modified Permutation Matrix Selection Model 123
6.3 The Quadratic Assignment Problem 127

Appendix A--An Introduction to Dynamical Systems

A.1 Elements of Two-Dimensional Dynamical Systems 129
A.2 Elements of n-Dimensional Dynamical Systems 132
A.3 The Relation Between Difference and Differential Equations 133
A.4 The Concept of Stability 138
A.5 Limit Cycles 140
A.6 Lyapunov Theory 142
A.7 The Linearization Theorem 146
A.8 The Stability of Linear Systems 147

Appendix B--Simulation of Dynamical Systems with Spreadsheets

Appendix B--Simulation of Dynamical Systems with Spreadsheets 151

References 160

Index of Key Words 163

Epilog 166

PREFACE

In mathematics there are limits, speed limits of a sort, on how many computational steps are required to solve certain problems. The theory of computational complexity deals with such limits, in particular whether solving an n-dimensional version of a particular problem can be accomplished with, say, n^2 steps or will inevitably require 2^n steps. Such a bound, together with a physical limit on computational speed in a machine, could be used to establish a speed limit for a particular problem.

But there is nothing in the theory of computational complexity which precludes the possibility of constructing analog devices that solve such problems faster. It is a general goal of neural network researchers to circumvent the inherent limits of serial computation.

As an example of an n-dimensional problem, one might wish to order n distinct numbers between 0 and 1. One could simply write all n! ways to list the numbers and test each list for the increasing property. There are much more efficient ways to solve this problem; in fact, the number of steps required by the best sorting algorithm applied to this problem is proportional to n ln n .

As explained by A. K. Dewdney in the June 84 and June 85 issues of *Scientific American*, the n numbers could also be ordered using an analog device. According to Dewdney, one could cut lengths of uncooked spaghetti in proportion to the numbers and then simply stand the spaghetti in one's hand on the surface of a table. The degree to which computational complexity speed limits apply to analog devices is not completely clear, but the point is that analog computation offers some promise of the efficient solution of certain problems. In fact, electronic computers storing numbers as voltages across capacitors have a long history and are currently used as dedicated devices for solving planetary motion problems. Certainly ordering n = 100 numbers with lengths of spaghetti would be easier than examining 100! lists of 100 numbers.

The purpose of this book is to explore some uses of one type of analog computer, the (artificial) neural network.

There are two basic types of neural network. In the *composition function neural network*, an input vector is fed to several nonlinear functions of a certain type. Then the values of those functions are fed to other functions (all the functions are arranged in *layers*), and so on until a final function is evaluated. The final function might yield a "yes" or "no" answer to a problem, or a series of forces to apply to a robot arm, or some other decision. The layers of intermediate functions would depend upon certain *weight parameters*; the *weights* are determined by *training* the neural network on a mass of historical data plus known decisions. Thereafter the neural network could be fed original input data and be expected to give new decisions of a high quality.

This book is concerned entirely with the second type, namely the *dynamical system neural network*. Here an input vector becomes an initial state of a *dynamical system*. The system iterates with time and converges to a state which can be interpreted as a decision. Much of this book is devoted to a neural network method for recognizing binary strings which have been corrupted by noise in transmission (recognizing input words despite some spelling errors, one might say). It appears that an electronic analog of the mathematical error-correcting decoder would correct errors faster and more reliably than conventional algebraic decoders.

What is the future of neural network research? Much interest in neural networks has historically been due to the hope of understanding the underlying mechanisms of human cognition. This is not a realistic goal as far as I am concerned. Nonetheless, there are good medical reasons for promoting basic research in artificial neural networks. The development of devices designed to serve patients externally (intelligent robotic patient care) depends upon the development of adaptive control. Patients with traumatic injury to or degenerative disease of the central nervous system would likely find the services of an intelligent robot especially valuable. Also, the general field of replacing or supplementing body components associated with manipulation, mobility, vision, hearing, speech, hormonal stability, or other biodynamical systems will also develop in step with intelligent control theory. Finally, it seems possible that some type of implanted neural network circuit might be used someday to replace lost cognitive capacity or to enhance normal cognitive capacity. This would be a fundamental step in human history.

Chapter 0--The Neural Network Approach to Problem Solving

0.1 Defining a Neural Network

In the language of computing, a neural network model is a dynamical system which serves as a fully parallel analog computer to mimic some aspect of cognition. As such, a neural network model simulates image recognition, learning, limb control, or some other function of a central nervous system.

In this book, a neural network model for image recognition functions in the following manner. Input by sensors becomes a vector in n-dimensional space. The model regards this as an initial state of a trajectory. The initial state is then processed by loop iteration of vector-valued *system functions*. Recognition is the subsequent convergence of the state vector to the nearest of the stored attractors. The stored attractors are vectors, binomial n-strings with components ±1 (each naturally equivalent to an n-string with components 0 and 1).

If the system functions have parameters that are somehow adjusted to fit known input-answer data, then adjusting the parameters is called *training*. An image recognition model is said to have *content addressable memory* because partial knowledge of a memory leads to the complete retrieval of the memory through system dynamics. Neural network recognition is formally equivalent to the general concept of error correction of binary code by means of an analog dynamical system, much in contrast to the use of serial algebraic algorithms in the conventional theory of error-correcting codes.

Alternatively, neural networks can be specified implicitly as having attractors with certain properties and thus solutions (attractors with those properties) can be discovered by repeated simulations with random initial conditions. It is possible to specify a neural network model which is effective in placing m rooks in nonthreatening positions on an m×m chessboard, an m^2-dimensional problem with permutation matrices as solutions. It is also possible to favor some feasible solutions over others by modifying certain parameters in the associated additive neural network model. For example, there might be a benefit (inverse of cost) associated with each of the squares of an mxm chessboard and one might wish to find a placement of m nonthreatening rooks with high total benefit.

In the theory of control of a limb or other physical plant, the plant state is regarded as the initial condition of a neural network with convergence to a memory corresponding to choice of a particular control measure. A sequence of control measures serves to drive the plant to a desired state in space or spacetime. Frequent cycles of sampling, neural network convergence to a memory, and control measure activation, all with training, would amount to a type of control. There are many neural network models that can learn from experimental data to control a system. For example, neural networks have been trained to adjust the forces at joints of robotic limbs to compensate for the flexing of beams and joints under load. The neural network approach is robust in the sense that one does not know or even attempt to know the exact nature of uncontrolled system dynamics; one tries to design the control so that it will function effectively regardless of the details of those dynamics.

Thus neural network dynamical systems are in principle capable of solving a variety of finite choice problems from coding theory, image recognition, combinatorics, operations research, and control theory. The general strategy is to design an analog computer in the form of an n-dimensional dynamical system with solutions to a problem that somehow represented as constant attractor trajectories (stable equilibria) or cyclic attractor trajectories (limit cycles).

A general goal of neural network research is the implementation of a design strategy as an electronic device for the purpose of fast analog computation. This transition has been studied by many researchers (see C. Mead [M] for an introduction). For certain problems a neural network might arrive faster (if implemented as a dedicated chip) at better solutions than would any conventional algorithm. In particular, for some classes of problems it might be possible to circumvent in a practical sense the limits of serial computational speed imposed by the theory of computational complexity.

0.2 Neural Networks as Dynamical Systems

In mathematical terms, the neural network models developed in this book are *dynamical systems* which solve finite selection problems arising in image recognition (especially coding theory), operations research, combinatorics, and control theory. In each such problem, answers are represented as particular binary n-strings each component of which is ± 1 or, equivalently, 0 or 1. Thus answers to selection problems are thought of as vertices of the n-cube, the n^2 vertices of which are the points in n-dimensional space with coordinates ± 1. The job of the neural network is to generate a series of points in n-space called a *trajectory* <u>from</u> some *initial state* (at time zero) in n-space <u>to</u> an answer vertex. Given a vertex of the n-cube to which some trajectories converge, the set of initial states of all such trajectories is called the *attractor region* of the vertex.

In the language of dynamical systems theory, a mathematical neural network model is an n-dimensional difference equation or differential equation *dynamical system* that accounts for the dynamics of n *neurons*. Each neuron is mathematically defined by its state x_i (a real number) and its associated output *gain function* $g_i = g_i(x_i)$. For each i = 1,2,...,n there must exist a positive constant $\varepsilon_i < .5$ with $g_i(x_i) = 0$ for $x_i \leq -\varepsilon_i$ and $g_i(x_i) = 1$ for $x_i \geq \varepsilon_i$; furthermore, $g_i(x_i)$ must be nondecreasing. A gain function is typically a ramp or sigmoidal function. An example of a gain function appears in Fig. 0.1.

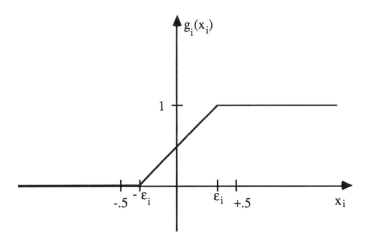

Figure 0.1. A gain function which might be used in a neural network model. Here ε_i is a parameter which, as it turns out, can be modified to alter the number and types of attractors of the model. Note the positive slope of the gain function *in transition* (where $|x_i| < \varepsilon_i$). In general the *gain* of a gain function is its slope at $x_i = 0$, here $(2\varepsilon_i)^{-1}$.

The time rate of change of each x_i is determined by a *system function* dependent on x_i itself and all the outputs $\{g_j(x_j)\}$. Any such system is just a formal version of the update formula

new = old +[rate of change]Δt

or the differential equation limit of this formula as $\Delta t \to 0$.

In this book we shall focus on a limited version of the dynamical system format. That is, the time rate of change of x_i will be determined by an Euler difference equation of the form

$$x_i(t+\Delta t) = x_i(t) + [-k_i x_i(t) + p_i(g(x(t)))]\Delta t \qquad (0.1)$$

or the corresponding differential equation limit of this formula as $\Delta t \to 0$,

$$dx_i/dt = -k_i x_i + p_i(g(x)) \qquad (0.2)$$

Here each k_i is a positive constant and each p_i is a multinomial function of the n variables $g_1(x(t))$, $g_2(x(t))$,...,$g_n(x(t))$ which behaves mathematically well enough so that trajectories for (0.1) or (0.2) exist and are unique. In this book, type (0.1) is used with a Δt sufficiently small to simulate the generic trajectories of (0.2). (Other neural network models are inherently of the type (0.1), always use $\Delta t = 1$, and have as state space only the vertices of the n-cube, not all of n-dimensional space.)

Our aim in solving problems with neural networks is to build an n-dimensional dynamical system with solutions of a problem represented as constant attractor trajectories (stable equilibria) or, as in Chapter 4, cyclic attractor trajectories (limit cycles). After rescaling, each constant attractor trajectory can be one of the 2^n points in n-dimensional space with coordinates ± 1. In geometric terms, the x values of the neural network model change with time and thus describe trajectories in n-dimensional space that asymptotically approach some of these points. The initial state of a trajectory must be determined by input data. In summary, one tries to couch possible answers to a decision problem in terms of such binomial n-strings and then specify the right side, the *system functions*, of a dynamical system (0.1) or (0.2) to generate the desired decision process as system trajectories.

0.3 Additive and High Order Models.

In *additive neural network models*, each p_i in (0.1) or (0.2) is a linear function of the components of g. (Additive models and many other neural network ideas can be found in the survey article by S. Grossberg [G].) *High order neural network* models have emerged recently as interesting alternatives to earlier additive models. In high order models, as devised by Y. Lee, G. Doolen, H. Chen, G. Sun, T. Maxwell, H. Lee, C.L. Giles, and others [GM], [LD...], [PPH], each p_i is a multinomial of components of g. In the high order models in this book, we will use as p_i multinomials of a special type: sums of products of g components, each product of the form

$$g_1{}^{e_1} g_2{}^{e_2} ... g_n{}^{e_n}$$

where each exponent e_i is either 0 or 1. This turns out to be equivalent (as explained in Chapter 2) to using sums of products of n terms each, the jth multiplicand being either g_j or $1-g_j$.

For example, in a five-dimensional model we might use a system function $p_1(g)$ given in expanded form by

$$p_1(g) = -g_2g_4+g_1g_2g_4+g_1g_3g_5+g_2g_4g_5+g_2g_3g_4-g_1g_2g_3g_5-g_1g_3g_4g_5$$
$$-g_1g_2g_3g_4-g_2g_3g_4g_5-g_1g_2g_4g_5+2g_1g_2g_3g_4g_5 \qquad (0.3)$$

or equivalently as a product of g_i and $1-g_i$ multiplicands

$$p_1(g) = g_1(1-g_2)g_3(1-g_4)g_5 - (1-g_1)g_2(1-g_3)g_4(1-g_5) \qquad (0.4)$$

There are two notable low order types of $p_i(g)$. A *linear* p_i function is of course a sum

$$\sum_{j=1}^{n} T_{ij} g_j$$

where T_{ij} are real constants comprising an nxn matrix. In a *quadratic* p_i

function each summand is expressed as a multinomial with two exponents in each product being 1, the rest being 0. Thus a quadratic p_i is given by a sum

$$\sum_{j,\,k=1}^{n} T_{ijk}\, g_j g_k$$

where $\{T_{ijk}\}$ can be thought of as a collection of n matrices, each nxn.

However, for general $p_i(g)$, if each g component is 0 or 1, then the form represented by (0.4) is simpler in that <u>at most one product is nonzero</u>. One can see immediately from (0.4) that two combinations of 0 or 1 values for $\{g_i\}$ will make $p_1(g) = 1$ (namely, $g = (1,0,1,0,1)$ and $g = (0,1,0,1,0)$. For all 30 other possible values for g, $p_1(g) = 0$.

0.4 Examples

One additive model (each p_i linear) in Chapter 5 is effective in placing m rooks in nonthreatening positions on an mxm chessboard. Here $n = m^2$ so we have an m^2-dimensional problem with mxm permutation matrices (again regarded as certain of the vertices of the m^2-cube) as solutions. It is also possible to favor some feasible solutions over others by modifying certain parameters in the associated additive neural network model. How good solutions can be found using a neural network model is also explained in Chapter 5.

In Chapter 1 certain models with quadratic $\{p_i\}$ and perturbation terms are used to solve *set selection problems*. In a set selection problem one is given a collection of subsets of a set with n elements and one wants to find yet another subset with certain intersection properties with respect to the given subsets. For example, the placement of n nonthreatening queens on an nxn chessboard is a set selection problem. A solution set must intersect every row subset exactly once, every column subset exactly once, and every diagonal subset at most once.

We use the term *infeasible* to describe a constant trajectory (steady state) for a neural network with at least one component of g *in transition*, that is, $0 < g_i < 1$. Generally an infeasible constant trajectory cannot be interpreted as a solution to a selection problem; half a rook on a square has no meaning. A fundamental difficulty with models using linear or quadratic p_i

functions is that they generally have many infeasible constant attractors. They also might have cyclic trajectories as attractors which cannot be interpreted as solutions.

The article [J] contains a specific high order neural network design, the *memory model*, in which all p_i are of the type represented in (0.4) with all coefficients ±1. The memory model can store using n neurons any number M, $1 \leq M \leq 2^n$, of any (or even all) of the binomial n-strings; in a schematic representation the model requires only $5n+M(1+2n)$ edges. Each stored n-string represents a memory as a constant attractor trajectory of an n-dimensional differential equation dynamical system. With sufficiently high gains (slope of gain function at $x_i = 0$), the only stable attractors are the memories. Thus the memory model amounts to a solution of a version of the fundamental memory problem posed in the neural network literature. A detailed presentation of the memory model appears in Chapter 3.

The memory model can be used in error-correcting decoding of any binary string code. Throughout the book use is made of the association

$$x_i = 1 \quad \Leftrightarrow \quad g_i = 1$$
$$x_i = -1 \quad \Leftrightarrow \quad g_i = 0$$

Thus a memory is regarded as both a point x in n-space with ±1 components or as a (0,1)-string m_ι, $\iota = 1,2,...,M$.

One advantage of the neural network approach over conventional serial implementaion of algorithms is that noisy code arriving in nonbinary form can be corrected without first rounding signal digits to binary form. This seems to make neural network decoding inherently more accurate than serial decoding.

However, error correction of binary codes is actually only a formal application of a much more general notion, *content addressable memory*. One wants to store memories as binary strings and one wants a system which, given partial information (meaning an initial state close to one of the memories) will go towards the closest memory. When the model arrives in a small neighborhood of a memory, it must then send a signal indicating that a particular memory has been reached. Thus some pattern recognition problems can be couched in terms of error-correcting code terminology.

In the language of the memory model neural network, each memory can correspond to the activation of a control: if the neural network is on memory

m_1, then control 1 is put into effect. The state of the system is sampled and the state data are translated into initial conditions for the neural network. The neural network then converges quickly to a memory and the associated control is activated. This process of sampling to control activation is repeated at a sufficiently high rate to maintain control. Shaping the basins of attraction of the memories in order to drive a system to a desired system state or desired time-dependent sequence of states is a version of learning.

0.5 The Link with Neuroscience

There is also a link between neuroscience and neural network models, but the significance of the link is not clear at present. One can only say that neural network models are indirectly inspired by neural structures, as were, for example, early aircraft by birds. Just as few aircraft today use flapping wings, mathematical neural network models differ fundamentally from natural neural networks. Most striking is the fact that natural neurons have a myriad of different effects on and responses to each other, much in contrast to the simple model of artificial neuron as amplifier.

It is likely to be quite some time before mathematicians and engineers build machines that process images, control limbs, or even reason as effectively as does, say, the nervous system of a chicken. But when they do, synthetic neural networks will prove vastly superior in many ways.

Chapter 1--Neural Networks as Dynamical Systems

1.1 General Neural Network Models

The activity levels of n neurons are represented by a point x in n-dimensional state space, that is, the space consisting of n-tuples $x = (x_1, x_2, ... , x_n)$ of real numbers. Define a *gain function* $g_i = g_i(x_i)$ by $g_i = 0$ if $x_i \leq -\varepsilon_i$; $g_i = 1$ if $x_i \geq \varepsilon_i$; and g_i is continuously differentiable and increasing for $|x_i| < \varepsilon_i$. We call $g_i'(0)$ the *gain* of g_i. For example, g_i might be *piecewise linear* (see Fig. 0.1), that is, a *ramp function*, with $g_i(x_i) = (x_i + \varepsilon_i)/(2\varepsilon_i)$ for $|x_i| < \varepsilon_i$.

Suppose for each $i = 1, 2, ... , n$ that p_i is a continuously differentiable function of n-space and k_i is a positive constant. For our purposes, a *neural network model* is a dynamical system

$$dx_i/dt = - k_i x_i + p_i(g(x)) \qquad\qquad (1.1)$$

Actually, (1.1) is 3^n dynamical systems in 3^n regions of state space, the regions in which each x_i satisfies $x_i \leq -\varepsilon_i$, $-\varepsilon_i < x_i < +\varepsilon_i$, or $x_i \geq +\varepsilon_i$. In each such region *trajectories* for (1.1) exist, are unique, and, as we shall see, generally extend forward in time to the boundary of the region or to an attractor in the region. Any vector-valued function of time which is made up of such trajectories is regarded as a *neural network trajectory* for (1.1).

Let us define the *transition zone* for (1.1) to be the set T_ε of points x in n-space with at least one component x_i satisfying $|x_i| < \varepsilon_i$. Clearly outside T_ε each $g_i(x_i)$ is constant and so each $p_i(g(x))$ is constant. It is also clear that the complement of T_ε in n-space consists of 2^n (closed) components, each one naturally being a subset of an orthant of n-space; we refer to each such component as an *orthant relative to T_ε*. Trajectories for (1.1) in an orthant O relative to T_ε are simply pieces of trajectories of the linear (constant coefficient) dynamical system $dx_i/dt = k_i(-x_i + c_i)$ where c_i is the constant $k_i^{-1} p_i(g(x))$; $c = (c_1, c_2, ...,c_n)$ itself might or might not lie in O.

In all that follows we assume that $c_i \neq \pm \epsilon_i$ for all orthants and all i. A given version of (1.1) which did not meet this criterion could be adjusted by reducing ϵ_i values as needed (c_i values would not be affected by such a change). A consequence of this assumption is that all trajectories which start in an orthant relative to T_ϵ and subsequently meet the boundary of that orthant do so transversally (not tangentially). This assumption avoids mathematical difficulties associated with trajectories lying in the boundary of an orthant relative to T_ϵ . Therefore, a trajectory starting in the boundary of T_ϵ must after one time step leave that boundary and must at each time then and thereafter use the system functions of exactly one of the 3^n regions.

To rephrase the introduction, the general purpose of a neural network model is to generate trajectories in n-dimensional space which asymptotically approach some of the constant attractor trajectories in the orthants relative to T_ϵ. There can be any number M of such attractors, $1 \leq M \leq 2^n$. The right side of (1.1) must somehow reflect a decision process as a trajectory, and the initial state of the trajectory must somehow be determined by input data.

To illustrate the above, let us consider the two-dimensional neural network model

$$dx_1/dt = -x_1 + g_1(1-g_2) - (1-g_1)g_2 \tag{1.2}$$
$$dx_2/dt = -x_2 - g_1(1-g_2) + (1-g_1)g_2$$

where each gain function is a ramp function with $\epsilon_1 = \epsilon_2 = .1$.

We note that (1.2) could be algebraically simplified to

$$dx_1/dt = -x_1 + g_1 - g_2$$
$$dx_2/dt = -x_2 - g_1 + g_2$$

However, doing so obscures the simple fact that the $p_i(g)$ are switched on (have nonzero value) in the (-,+)-orthant and the (+,-)-orthant (second and fourth quadrants) and switched off (have value 0) in the (+,+)-orthant and the (-,-)-orthant. Furthermore, the generalization of (1.2), which we shall use below, is represented by the form in (1.2). Thus we regard (1.2) as a high order model even though it is also a linear model because n is only 2.

Typical trajectories for (1.2) are shown in Fig. 1.1.

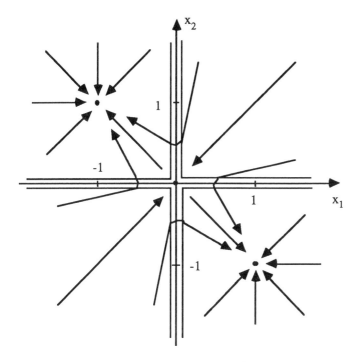

Figure 1.1. Typical trajectories for a memory model in two-dimensional space with two memories. The transition zone for the model is the "+" shaped region enclosing the axes.

Let us recall (from Appendix A) a few dynamical systems concepts. A *constant trajectory* is a constant solution $x(t) = x(0)$ of (1.1), that is, a vector $x(0)$ for which each component of the right side of (1.1) is 0.

The *distance* $d(x,y)$ in n-dimensional space between vectors x and y is defined by

$$d(x,y) = \left(\sum_{i=1}^{n} (x_i - y_i)^2 \right)^{.5}$$

A constant trajectory is called *stable* if two conditions are satisfied:

first, there must be a positive number ε such that every trajectory starting within ε of $x(0)$ must asymptotically approach $x(0)$;

second, for any positive number ε there must be a positive number $\delta(\varepsilon)$ such that a trajectory can be guaranteed to stay within ε of $x(0)$ just by requiring that it start within $\delta(\varepsilon)$ of $x(0)$.

The set of all points which can serve as initial states for trajectories that asymptotically approach a stable trajectory is called the *attractor region* or *region of attraction* of the stable trajectory.

The system (1.2) happens to have three constant trajectories: $(-1,1)$, $(0,0)$, and $(1,-1)$. Except for the two special trajectories that asymptotically approach the unstable constant trajectory $(0,0)$, nonconstant trajectories asymptotically approach either the constant trajectory $(-1, 1)$ (with g vector $(0,1)$) or the constant trajectory $(1, -1)$ (with g vector $(1,0)$). Thus (1.2) is an almost trivial system for generally deciding which of two memorized states is closer to a given state (used as the initial state of a trajectory). To put things another way, the system partitions the (x_1,x_2)-plane into two attractor regions separated by the line $x_2 = x_1$. Other decision problems treatable with the neural network approach are not so simple.

A *limit cycle* is a simple closed curve in n-dimensional space with the following properties. First, no constant trajectories are contained in the limit cycle. Second, any trajectory which starts at a point in the limit cycle must stay in the limit cycle thereafter. Third, there must be a positive number ε such that every trajectory which starts within ε of the limit cycle must asymptotically approach the limit cycle. Fourth, for any positive number ε, there must be a positive number $\delta(\varepsilon)$ such that a trajectory can be guaranteed to stay within ε of the limit cycle just by requiring that it start within $\delta(\varepsilon)$ of the limit cycle.

We note that the speed of a trajectory on the limit cycle (the magnitude $|dx/dt|$ of the vector dx/dt) is a continuous function on a compact set. Since speed cannot be zero, it follows that it must be bounded as

$$0 < b < |dx/dt| < B < \infty$$

by constants b, B.

To illustrate the concept of limit cycle we define a second two-dimensional neural network model by the equations

$$dx_1/dt = -x_1 -g_1g_2 -2(1-g_1)g_2+(1-g_1)(1-g_2)+2g_1(1-g_2) \tag{1.3}$$
$$dx_2/dt = -x_2 +2g_1g_2 -(1-g_1)g_2-2(1-g_1)(1-g_2)+g_1(1-g_2)$$

where the gain function g is a ramp function as in the previous example with ε = ε_1 = ε_2. The *linear approximation matrix* of a dynamical system at a constant trajectory is the Jacobian

$$L = \left(\partial(dx_i/dt)/\partial x_j\right)_{x=\text{constant trajectory}}$$

For (1.3) at the constant trajectory (0,0) we have

$$L = \begin{pmatrix} -1+g' & -3g' \\ 3g' & -1+g' \end{pmatrix}$$

where $g' = 1/(2\varepsilon)$. According to the Linearization Theorem (see Appendix A or [W, p. 125-130]), the stability of trajectories of nonlinear systems can be tested by analyzing a linear approximation of the system about the trajectory, namely,

$$dy/dt = L\, y$$

where $y_i = x_i - x_{i\,\text{constant trajectory}}$. In this case the real parts of the eigenvalues of L are both $-1+g'$. It follows from the properties of linear systems (see Appendix A or [W, p. 61]) that (0,0) is unstable if $g' > 1$, that is, $.5 > \varepsilon$.

    With some effort, one can prove that the trajectories of (1.3) with $\varepsilon = .1$ are as shown in Fig. 1.2. The origin is the only constant trajectory, the closed trajectory is a limit cycle, and all other trajectories asymptotically approach the limit cycle. If we were to use smaller values for ε, then the limit cycle would come closer to the square with vertices $(\pm1,0)$ and $(0,\pm1)$.

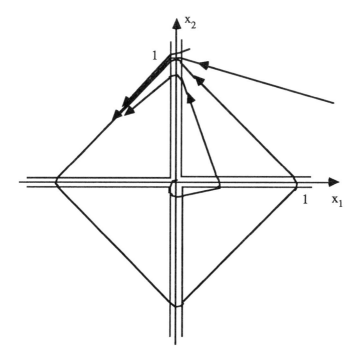

Figure 1.2. A neural network model with a limit cycle. The transition zone for the model is the "+" shaped region enclosing the axes.

While in the $(+,+)$-orthant relative to T_ε (where $g_1 = g_2 = 1$), the system (1.3) amounts to the linear system

$dx_1/dt = -x_1 - 1$
$dx_2/dt = -x_2 + 2$

Thus, while in the $(+,+)$-orthant, trajectories for (1.3) pursue $(-1,2)$. As the trajectory crosses from the $(+,+)$-orthant to the $(-,+)$-orthant, g changes from $(1,1)$ to $(0,1)$. Similarly, while in the $(-,+)$-orthant, trajectories for (1.3) pursue $(-2,-1)$. As the trajectory crosses from the $(-,+)$-orthant to the $(-,-)$-orthant, g changes from $(0,1)$ to $(0,0)$. So the changes in g continue around the cycle. This cycle tours orthants of 2-space so that g values are $...\rightarrow(+,+)\rightarrow(-,+)\rightarrow(-,-)\rightarrow(+,-)\rightarrow(+,+)\rightarrow...$ As we shall see in Chapter 4, it is possible to generalize this design to obtain n-dimensional neural network models with various limit cycles. Thus it is possible to *memorize* (specify as attractors) cyclic patterns as well as constant patterns in the high order neural network format.

1.2 General Features of Neural Network Dynamics

We have seen that simple neural network models can exhibit multiple constant trajectories and limit cycles. However, the dynamics of neural networks far from the origin are quite restricted.

An *attractor region* for a dynamical system is a set of points in a state space of finite diameter which every trajectory ultimately enters and does not exit. It is easy to prove (see [JvdD]) the following facts about (1) (see [LMP] for other models):

THEOREM 1.1. Every neural network model

$$dx_i/dt = -k_i x_i + p_i(g(x)) \tag{1.1}$$

has a finite attractor region.

PROOF. Since for each i we have $0 \le g_i(x_i) \le 1$, there is a positive constant P with $-P \le k_i^{-1}p_i(g(x)) \le P$. Thus for each i we have $k_i(-x_i - P) \le dx_i/dt \le k_i(-x_i + P)$. It follows that if $x_i > 0$ then

$$x_i \, dx_i/dt \le k_i(-x_i^2 + Px_i)$$

while $x_i < 0$ implies

$$x_i \, dx_i/dt \le k_i(-x_i^2 - Px_i)$$

Consider the 2^n ellipsoids centered at $(\pm_1 P/2, \pm_2 P/2, ..., \pm_n P/2)$ on which Σ $k_i(x_i^2 \pm_i Px_i)$ is zero. Let B be the n-ball centered at the origin and supported by the 2^n ellipsoids. Let D(x) denote the distance from a point x to the origin of n-space. Clearly outside B, the derivative of D(x) is negative. For any $\delta > 0$ there is a ball B_δ of larger radius r_δ outside of which $dD/dt < -\delta$.

Suppose a trajectory for (1.1) lies initially at a point at a distance $D_0 > 0$ from B_δ. As time advances, the trajectory can take at most $(D_0 - r_\delta)/\delta$ time units before it must enter B_δ. Therefore B_δ is an attractor region for (1.1). QED

THEOREM 1.2. In each orthant O relative to T_ϵ either there is one stable constant trajectory c or no constant trajectories at all. In the former case, every trajectory that enters O thereafter asymptotically approaches c. In the latter case, every trajectory that starts in O must, while in O, asymptotically approach some point outside O and so must exit O after a finite time interval.

PROOF. Let O be an orthant relative to T_ϵ. Let c_i be the (constant) value of $k_i^{-1}p_i(g(x))$ in O. Thus the only possible constant trajectory for (1.1) in O is c $= (c_1, c_2,..., c_n)$.

The system (1.1) is, of course, a linear system in the interior of O given by

$$dx_i/dt = -k_i(x_i - c_i)$$

If a constant trajectory exists in O, then relative to that point the matrix of the system is a diagonal matrix with negative entries (which are for such a matrix the eigenvalues). Hence the constant trajectory is asymptotically stable. QED

General trajectories in O are given explicitly by

$$x_i(t) = c_i + (x_i(0)-c_i)\exp[-k_i t]$$

Generally speaking, stable constant trajectories for (1.1) also exist in T_ϵ ; such constant trajectories are called *infeasible*. Infeasible stable constant trajectories in T_ϵ are undesirable features of neural network models. To put it another way, we wish to design neural network models so that in all constant attractor trajectories each neuron is either *on* ($g_i = 1$) or *off* ($g_i = 0$), and not in an intermediate state.

1.3 Set Selection Problems

Neural network models can be used to solve a variety of set selection problems and set domination problems. Such applications also serve to introduce the general dynamical properties of neural networks.

Let us denote by S the finite set with n elements. Let | | denote the cardinality operator, so that $|S| = n$. Suppose with S are associated nonempty subsets $S_1, S_2,..., S_p$. Our first *set selection problem* entails the selection of yet another subset of S, an *answer set A*, satisfying :

$$|A \cap S_k| = 1 \tag{1.4}$$

for k = 1,2,...,p.

As an example of a set selection problem, let us suppose S consists of the squares on an m×m chessboard, so $n = m^2$. As subsets $\{S_k\}$ we take the squares in the m rows and the m columns of the board. We might then seek a subset A so that exactly one element of A is in each row subset and exactly one element of A is in each column subset. This would amount to the placement of m nonthreatening rooks on the chessboard or the selection of an m×m permutation matrix.

In n-dimensional space, that is, the space consisting of n-tuples x = $(x_1, x_2, \dots, x_n)$ of real numbers, any solution set A can be thought of as a vertex of the n-cube in the following way. All the vertices of the n-cube can be naturally identified with the 2^n points $\{(\pm 1, \pm 1, \dots, \pm 1)\}$. If element i is in A, then let the ith component of the vertex n-tuple be +1; otherwise let the ith component be -1.

We shall specify dynamical system equations in terms of certain set selection problems so that trajectories starting at randomly selected initial points in n-space asymptotically approach with time a vertex of the n-cube (a feasible constant trajectory) which is equivalent to a solution set. The approach to be used amounts to an extension of the ideas of Cohen and Grossberg [CG] on content-addressable memory, Tank and Hopfield [TH] on neural optimization networks, and Page and Tagliarini [PT] on general algorithm development for neural networks.

Suppose in a set selection problem that element i is contained in a number n_i of the subsets $\{S_j\}$ and each $n_i \geq 1$. Let us define a neural network model for this set selection problem by

$$dx_i/dt = -x_i + 1 -(2/n_i) \sum_{\substack{S_j \\ i \in S_j}} \sum_{\substack{k \in S_j \\ k \neq i}} g_k(x_k) \tag{1.5}$$

where the first sum is over all subsets S_j containing element i and the second sum is over all $k \in S_j$ with $k \neq i$.

Any constant trajectory for (1.5) must satisfy

$$x_i = +1 -(2/n_i) \sum_{\substack{S_j \\ i \in S_j}} \sum_{\substack{k \in S_j \\ k \neq i}} g_k(x_k) \tag{1.6}$$

We proceed to characterize the constant trajectories of (1.5).

THEOREM 1.3. Every answer set for the above set selection problem (1.4) corresponds to a stable constant trajectory x with each $x_i = \pm 1$ for the neural network (1.5). Also, every constant trajectory for the neural network with every $x_i = \pm 1$ corresponds to an answer set for the set selection problem.

PROOF. Suppose A is an answer set. If neuron i is in the answer set, define the ith component of an n-vector by $x_i = +1$; otherwise define $x_i = -1$. We proceed to show that x satisfies (1.6), that is, x is a constant trajectory for (1.5). Suppose $x_j = 1$. Then $g_j = 1$ and for all $k \neq i$ in all the subsets containing element i, $g_k = 0$. Thus the right side of the ith equation in (1.5) adds up to zero. Suppose $x_j = -1$. Then $g_j = 0$ and for exactly one $k \neq i$ in each of the subsets containing neuron i, $g_k = 1$; other g values in such subsets are all 0. Thus the right side of the jth equation in (1.5) again adds up to zero. Thus x is a constant trajectory for (1.5) in an orthant relative to T_ε and so, by Theorem 1.2, is stable.

Now suppose x is a constant trajectory for (1.5) with every $x_i = \pm 1$. Then each component of g is either 0 or 1. Consider the ith equation in (1.6). If $x_i = 1$, then all g_k in the sum must be 0, that is, all other neurons in all the subsets containing i must be off. At most one neuron can be on in each subset. Alternatively, if $x_i = -1$, then n_i of the g_k in the sum must be 1; since at most one neuron can be on in each subset, exactly one must be on in each subset. Thus x corresponds to an answer set.   QED

Theorem 1.3 establishes a correspondence between answer sets for the set selection problem (1.4) and constant attractor trajectories for (1.5) which have $x_i = \pm 1$. However, (1.5) does not solve (1.4) in the sense that other stable constant attractors that do not correspond to answer sets might exist for (1.5).

Suppose we are given a neural network model (1.1), perhaps a set selection model. Assume the model has a constant trajectory x with each g component 0, .5, or 1, that is, each component of x satisfies $x_i < -\varepsilon_i$, $x_i = 0$, or $x_i > \varepsilon_i$. It follows that reducing some or all of the ε_i magnitudes leads to a new model which still has x as a constant trajectory. We call any such constant trajectory ε-invariant; such trajectories must be considered carefully in set selection problems.

The following is an example of a set selection model with ε-invariant constant trajectories. Suppose the set $\{1,2,3,4,5,6\}$ has subsets $\{1,3\}$, $\{2,3\}$, $\{3,4\}$, $\{4,5\}$, $\{4,6\}$. The associated version of (1.5) is

$$dx_1/dt = 1 - x_1 - 2g_3 \tag{1.6}$$
$$dx_2/dt = 1 - x_2 - 2g_3$$
$$dx_3/dt = 1 - x_3 - (2/3)(g_1+g_2+g_4)$$
$$dx_4/dt = 1 - x_4 - (2/3)(g_3+g_5+g_6)$$
$$dx_5/dt = 1 - x_5 - 2g_4$$
$$dx_6/dt = 1 - x_6 - 2g_4$$

There are two answer sets for the set selection problem, $A = \{1,2,4\}$ and $A = \{3,5,6\}$. The corresponding constant trajectories are $x = (1,1,-1,1,-1,-1)$ and $x = (-1,-1,1,-1,1,1)$. However, if $\varepsilon < 1/3$, then $(-1,-1,1/3,1/3,-1,-1)$ is a stable constant trajectory for (1.6) with g values $(0,0,1,1,0,0)$, that is, a stable constant trajectory not associated with an answer set.

It is also possible to specify set selection problems with each $n_i = 2$ and stable ε-invariant constant trajectories inside the transition zone. The following model has trajectories in T_ε which are stable for all positive $\varepsilon < .5$. Assume again that $n = 5$ and that A must intersect each of the subsets $\{1, 2\}$, $\{2, 3\}$, $\{3, 4\}$, and $\{4, 5\}$ exactly once. The answer sets are $A = \{1,3,5\}$ and $A = \{2,4\}$. The associated version of (1.5) is

$dx_1/dt = -x_1 +1 - 2g_2$ (1.7)

$dx_2/dt = -x_2 +1 -g_1-g_3$

$dx_3/dt = -x_3 +1 -g_2-g_4$

$dx_4/dt = -x_4 +1 -g_3-g_5$

$dx_5/dt = -x_5 +1 - 2g_4$

If each $\varepsilon_i < .5$, then the following are the constant trajectories for (1.7):

$x = (0, 0, 0, 0, 0)$	$g = (.5,.5,.5,.5,.5)$	(unstable)
$x = (1,-1, 1, -1, 1)$	$g = (1,0,1,0,1)$	(stable)
$x = (-1, 1, -1, 1, -1)$	$g = (0,1,0,1,0)$	(stable)
$x = (1,-.5, 0, .5, -1)$	$g = (1,0,.5,1,0)$	(stable)
$x = (-1, .5, 0,-.5, 1)$	$g = (0,1,.5,0,1)$	(stable)

Let us consider the eigenvalues of the linear approximation matrix:

$$L = \left(\partial(dx_i/dt)/\partial x_j\right)_{x=\text{constant trajectory}}$$

At both $(1,-.5, 0, .5, -1)$ and $(-1, .5, 0,-.5, 1)$, L for (1.7) is

$$L = \begin{pmatrix} -1 & 0 & 0 & 0 & 0 \\ 0 & -1 & -g' & 0 & 0 \\ 0 & 0 & -1 & 0 & 0 \\ 0 & 0 & -g' & -1 & 0 \\ 0 & 0 & 0 & 0 & -1 \end{pmatrix}$$

All of the eigenvalues of L are -1 . Thus L is a stable matrix and therefore $(1,-.5, 0, .5, -1)$ and $(-1, .5, 0,-.5, 1)$ are stable constant trajectories regardless of the value of g'.

Let us consider the related system

$dx_1/dt = .5(-x_1 +1) - g_2$

$dx_2/dt = -x_2 +1 -g_1-g_3$

$dx_3/dt = -x_3 +1 -g_2-g_4$

$dx_4/dt = -x_4 +1 -g_3-g_5$

$dx_5/dt = .5(-x_5 +1) - g_4$

This system has the same constant trajectories as (1.7) and meets the conditions of a theorem of Cohen and Grossberg [CG] because each p_i is some

$$\sum_{j=1}^{n} T_{ij}\, g_j$$

where T_{ij} is a symmetric matrix (called the *connection matrix* of the neural network). The linear approximation matrix at $(1,-.5, 0, .5, -1)$ and $(-1, .5, 0,-.5,\ 1)$ is

$$L = \begin{pmatrix} -.5 & 0 & 0 & 0 & 0 \\ 0 & -1 & -g' & 0 & 0 \\ 0 & 0 & -1 & 0 & 0 \\ 0 & 0 & -g' & -1 & 0 \\ 0 & 0 & 0 & 0 & -.5 \end{pmatrix}$$

It follows that both $(1,-.5, 0, .5, -1)$ and $(-1, .5, 0,-.5,\ 1)$ are stable regardless of the magnitude of g'.

According to the theorem, every trajectory must converge to some constant trajectory. However, two of the stable trajectories are infeasible and remain so as gain is increased arbitrarily. This example demonstrates that the theorem of Cohen and Grossberg applied to a Hopfield model does not preclude infeasible, stable constant trajectories.

Experiments with $\varepsilon_i = .1$ show that a trajectory for (1.7) starting at a random point in the n-cube ultimately approaches one of the feasible solutions in about 90% of simulations and otherwise approaches one of the infeasible solutions. While one can reduce the relative size of the attractor region of infeasible solutions by increasing gain, there is a computational cost: smaller Δt is needed to simulate trajectories through T_ε. Infeasible stable constant trajectories are general and undesirable features of additive neural network models.

1.4 Infeasible Constant Trajectories

We can obtain partial information on infeasible constant trajectories for the set selection model as follows.

THEOREM 1.4. Suppose a constant trajectory x for (1.5) has every g = 0 or 1, that is, every $x_i < -\varepsilon_i$ or $x_i > \varepsilon_i$. If neuron i is on, then the total number of other neurons which are on in all the subsets containing neuron i is less than $n_i/2$. In particular, the total number of neurons which can be on in any one subset S is less than $\max_{i \in S}\{1+n_i/2\}$. If neuron i is off, then the total number of other neurons which are on in all the subsets containing neuron i is greater than $n_i/2$.

PROOF. Suppose the statement is false in the sense that neuron i is on and the total number of other neurons which are on in the subsets containing neuron i exceeds $n_i/2$. Then the right side of (1.6) is negative, so neuron i is off, a contradiction. A subset S containing more than $\max_{i \in S}\{1+n_i/2\}$ neurons which are on leads to a similar contradiction.

Suppose the statement is false in the sense that neuron i is off and the total number of other neurons which are on in the subsets containing neuron i is less than $n_i/2$. Then the right side of (1.6) is positive, so neuron i is on, a contradiction.   QED

We will develop further results on infeasible constant trajectories in Chapter 6.

1.5 Another Set Selection Problem

Suppose now that each element $i \in S$ is contained in exactly two distinct subsets R_j and C_k (one of which might consist of the element i only). For example, adjacent blank squares in a grid like

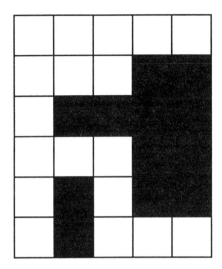

might be thought of as row segments $\{R_j\}$ and column segments $\{C_k\}$. We seek an answer set A with the following properties:

$$|A \cap R_j| \leq 1 \quad \text{for each } R_j \tag{1.8}$$
$$|A \cap C_k| \leq 1 \quad \text{for each } C_k$$

$$\bigcup_{|R_j \cap A|=1} R_j \quad \bigcup_{|C_k \cap A|=1} C_k = S$$

Thus the subsets which intersect A *cover* S. This is a version of a covering problem from combinatorics.

We shall analyze a related neural network model. First we define a *step gain function* in terms of a positive constant $\varepsilon < .5$ by $h(x) = 0$ for $x \leq 0$; $h(x) = 1$ for $x \geq \varepsilon$; and $h(x) = x/\varepsilon$ for $0 < x < \varepsilon$. Then we define a dynamical system by

$$dx_i/dt = 1 - x_i - 2 \left[\left(\sum_{\substack{j \in R_a \ j \neq i \\ i \in R_a}} g_j \right) + \left(\sum_{\substack{j \in C_b \ j \neq i \\ i \in C_b}} g_j \right) - h\left(\left(\sum_{\substack{j \in R_a \ j \neq i \\ i \in R_a}} g_j \right) \left(\sum_{\substack{j \in C_b \ j \neq i \\ i \in C_b}} g_j \right) \right) \right]$$

$$\tag{1.9}$$

THEOREM 1.5. Every answer set for the set selection problem (1.8) corresponds to a stable constant trajectory x with each $x_i = \pm 1$ for the neural network (1.9). Also, every constant trajectory for the neural network with every $|x_i| > \varepsilon$ corresponds to an answer set for the set selection problem and actually has $x_i = \pm 1$.

PROOF. Suppose A is an answer set. If neuron i is in the answer set, define the ith component of an n-vector by $x_i = +1$; otherwise define $x_i = -1$. We proceed to show that x is a constant trajectory for (1.9). Suppose $x_j = 1$. Then $g_j = 1$ and for all $j \neq i$ in the two subsets R_a and C_b containing i, $g_j = 0$. Thus the right side of the jth equation in (1.9) adds up to zero. Suppose $x_j = -1$. Then $g_j = 0$. By the covering property, at least one and possibly two vertices in the union of R_a and C_b must have $g_k = 1$; other g values in such subsets are all 0 by the intersection property. In either case the right side of the ith equation in (1.9) again adds up to zero. Thus x is a constant trajectory for (1.9) in an orthant relative to T_ε and so, by Theorem 1.2, is stable.

Now suppose x is a constant trajectory for (1.9) with every $|x_i| > \varepsilon$. Thus each component of g is either 0 or 1 and the term in [ ] must be a nonnegative integer. Consider the ith equation in (1.9). If $x_i > \varepsilon$, then [ ] in (1.9) must be zero. Thus all g_j in both of the sums in [ ] must be 0, that is, all other neurons in the subsets R_a or C_b must be off. Suppose $x_i < -\varepsilon$ and two neurons j and k are on in R_a (so i,j,k are distinct). This leads to a contradiction in the dx_i/dt equation. Thus at most one neuron can be on in R_a. Similarly at most one neuron can be on in C_b. This establishes the intersection part of (1.8). Now suppose that no neuron is on in R_a or C_b. Then $x_i < -\varepsilon$ and the equation (1.9) for dx_i/dt yields $0 = 1 - x_i$, a contradiction. This establishes the covering part of (1.8). Thus x corresponds to an answer set.   QED

1.6 Set Selection Neural Networks with Perturbations

Let us consider terms of the type $[\Sigma g_j(x_j)-1]^2$ where the sum is over all j in a given subset S_k in a set selection problem. Each such term is nonnegative and is zero at any feasible constant trajectory. Simply adding a constant multiple of such a term to the right side of (1.5) would generally create new, unwanted constant trajectories. However, each such term could be multiplied by a randomly varying function $D_i(t)$ and added to dx_i/dt at each time step of a discrete version of (1.5) without changing the coefficients of a given feasible constant trajectory or, by Theorem 1.2, the stability of that trajectory in its orthant relative to T_ε. That is, we can change (1.5) to

$$dx_i/dt = -x_i + 1 - \frac{2}{n_i} \sum_{\substack{S_k \\ i \in S_k}} \sum_{\substack{j \in S_k \\ j \neq i}} g_k(x_k) + D_i(t) \sum_{\substack{S_k \\ i \in S_k}} \left[\sum_{j \in S_k} g_j(x_j) - 1 \right]^2$$

(1.10)

where the first and third sums are over all subsets S_k containing element i, the second sum is over all $j \in S_k$ with $j \neq i$, and the fourth sum is over all $j \in S_k$. Use of random $D_i(t)$ would tend to destabilize infeasible attractors.

To illustrate the idea in (1.10) let us add such a sum to each system equation in (1.7):

$$dx_1/dt = -x_1 + 1 - 2g_2 + D(t)(g_1+g_2-1)^2$$ (1.11)
$$dx_2/dt = -x_2 + 1 - g_1-g_3 + D(t)[(g_1+g_2-1)^2 + (g_2+g_3-1)^2]$$
$$dx_3/dt = -x_3 + 1 - g_2-g_4 + D(t)[(g_2+g_3-1)^2 + (g_3+g_4-1)^2]$$
$$dx_4/dt = -x_4 + 1 - g_3-g_5 + D(t)[(g_3+g_4-1)^2 + (g_4+g_5-1)^2]$$
$$dx_5/dt = -x_5 + 1 - 2g_4 + D(t)(g_4+g_5-1)^2$$

In a discrete version of (1.11), using Euler's method with $\Delta t = .1$, let us specify $D(t) = $ if($rand_1 > .5$, 1, -1) * if($rand_2 > .75$, 1, 0) * 10 where each rand is a uniformly random number between 0 and 1. Thus in one of four time steps on average, $D(t)$ is ± 10 (equally probable); otherwise $D(t) = 0$. Such an iteration scheme generally has the effect of perturbing a trajectory away from an infeasible stable constant trajectory or stable limit cycle. Computer experiments show trajectories starting at a randomly selected point inside the 5-cube with vertex components ± 1 converge to one of the two (of 32) feasible vertices within approximately fifty time steps on average.

We can also perturb a discrete version of (1.11) using Euler's method with $\Delta t = .1$, specifying $D(t) = \pm 10$ every fourth time step on average, 0 otherwise. Computer experiments show trajectories starting at a randomly selected point inside the 5-cube with vertex components ± 1 converge to one of the two feasible vertices within fifty time steps on average.

As a final example, let $n = 10$; let the subsets be $\{1, 2, 3\}$, $\{2, 3, 4, 5\}$,

$\{3, 5, 6, 9, 10\}$, $\{5, 9\}$, $\{1, 6, 7, 8\}$, $\{7, 8, 10\}$, and $\{7, 9\}$. The only solution set of the 1024 subsets is $\{2, 8, 9\}$. The Euler difference equation version of the 10-dimensional dynamical system constructed as in (1.8) with $\Delta t = .1$ and $D(t) = \pm 10$ in one fourth of the iterations, 0 otherwise, typically finds the solution in about 100 time steps.

1.7 Learning

Both the composition function model used by other authors and the dynamical system neural network model in this book are mathematical means of going from an input vector (with real or binomial values) to a decision (generally expressed as a binomial string). Within these models are function parameters (*weights*) which can be adjusted to affect the decision associated with an input. The general idea of adjusting parameters on the basis of testing the model is a form of *learning* or, from the external viewpoint, *training*. The whole point of using composition function models is the ease with which learning can be accomplished; such models are designed for learning from masses of historical data. One might also say that the whole point of using dynamical system models is the avoidance of learning by prespecifying what decisions should accompany random or sensory inputs. This prespecification can occur implicitly as with combinatorial problems, or explicitly, as with models choosing the nearest in a given list of binomial strings.

However, dynamical system models can also learn in several senses. First, the number and selection of binomial memories could be changed by including or deleting the feedback loops associated with memories. Second, the relative sizes of regions of attraction can be changed by altering such system parameters as ε_i. Third, parameters of auxilliary functions such as the step gain function h in (1.9) can be altered, thus affecting the sizes and shapes of attractor regions.

In solving set selection problems with perturbed neural networks, learning in yet another sense should be possible. The reduction of the number of time steps needed on average to reach an orthant containing a solution might be achieved by optimizing the frequency and magnitude of the perturbation factor $D(t)$ for a given problem. Optimizing $D(t)$ magnitudes amounts to learning how to solve such problems, and is in itself a dynamic procedure. This is reminiscent of a child learning to solve a jigsaw puzzle. He tries a piece, hoping for convergence, and learns that if the piece does not fit after a certain amount of effort, a new piece should then be tried. This theme is a feature of learning that is by no means restricted to children.

PROBLEMS

1. Consider the subsets $\{S_1, S_2,...,S_7\}$ of $S = \{1,2,...,10\}$ given by $\{1,2,3\}$, $\{2,3,4,5\}$, $\{3,5,6,9,10\}$, $\{5,9\}$, $\{1,6,7,8\}$, $\{7,8,10\}$, and $\{7,9\}$. Devise a set selection neural network with perturbation term which finds another subet A such that $|A \cap S_i| = 1$ for $i = 1,2,...,7$.

2. Sensors are to be place in blank squares in the grids shown below. If two blank squares are adjacent or if between two blank squares in the same row or column are only other blank squares, then we say each blank square is *visible* to the other. When a sensor is placed in a blank square we say it *covers* that square plus all squares visible to that square. For the grids shown below, devise a neural network which places sensors on blank squares so that: all blank squares are covered; and no two sensors are in the same row or column.

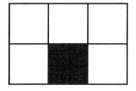

ANSWERS

1. The set selection model is

$$dx_1/dt = 1 - x_1 - (2/2)[g_2+g_3+g_6+g_7+g_8]$$
$$dx_2/dt = 1 - x_2 - (2/2)[g_1+g_3+g_3+g_4+g_5]$$
$$dx_3/dt = 1 - x_3 - (2/3)[g_1+g_2+g_2+g_4+g_5+g_5+g_6+g_9+g_{10}]$$
$$dx_4/dt = 1 - x_4 - (2/1)[g_2+g_3+g_5]$$
$$dx_5/dt = 1 - x_5 - (2/3)[g_2+g_3+g_4+g_3+g_6+g_9+g_{10}+g_9]$$
$$dx_6/dt = 1 - x_6 - (2/2)[g_3+g_5+g_9+g_{10}+g_1+g_7+g_8]$$
$$dx_7/dt = 1 - x_7 - (2/3)[g_1+g_6+g_8+g_8+g_{10}+g_9]$$
$$dx_8/dt = 1 - x_8 - (2/2)[g_1+g_6+g_7+g_7+g_{10}]$$
$$dx_9/dt = 1 - x_9 - (2/3)[g_3+g_5+g_6+g_{10}+g_5+g_7]$$
$$dx_{10}/dt = 1 - x_{10} - (2/2)[g_3+g_5+g_6+g_9+g_7+g_8]$$

At each time step we calculate the squares of the seven sums $(g_1+g_2+g_3-1)^2$, ..., $(g_7+g_9-1)^2$ corresponding to the seven subsets. Each of these squares is multiplied by 0 (probability p) or +1 (probability .5(1-p)) or -1 (probability .5(1-p)). This product is in turn added to the current equation for $x(t+\Delta t)$. It can be shown that with $\varepsilon = .1$ and $p = .75$ that convergence to an answer set typically takes 50 time steps.

2. This problem can be solved in the format of a spreadsheet (see the following pages). For the first grid, the constants Δt and ε are stored in cells B2 and E2. Initial values are stored in the B5..D6 block in the same pattern as in the grid. Values for $g_i(x_i)$ are stored in the B8..D9 block. For example, in B8 is entered the formula

=IF(B5<-$E$2,0,IF(B5>$E$2,1,(B5+$E$2)/(2*$E$2)))

This formula may be efficiently copied to other g cells. The "set sums R" block in B11.D12 is for each blank square just the sum of the covered squares in the same row. For example, in cell B11 is the formula

=B8+C8+D8

Similarly, set "sums C for columns" are entered. The "differences R" block is the R sum minus g_i itself. Hence in cell B17 the formula is

=B11-B8

The "g(diff)" block uses the piecewise linear gain function g which goes from 0 to 1 as its argument goes from 0 to e. This function is applied to the product (row sum - g_i)(col sum - g_i). Hence in cell B23 is the formula

=IF(b17*B20<0,0,IF(B17*B20>$E$2,1,B17*B20/$E$2))

The "next x" block is the time step itself. In cell B26 is entered the formula

=B5+(-B5+1-2*(B17+B20-B23))*$B$1

The "loop" block is just a copy of the "next x" values. Hence in cell B29 is the formula

=B26

One must allow for iterations by first clicking on "Iteration" in "Calculation" under "Options." Back up in cell B3 is the formula

=B27

When B3 is copied into the initial block B5..D6, the iteration loop is closed. In "Calculation" under "Options" one can specify the desired number of iterations.

The whole spreadsheet can be started at a random point with coordinates of magnitude < 1 as follows. Enter in cell B2 the formula

=2*(RAND()-.5)

Copying this cell into the initial block B5..D6 gives the dynamical system a random initial state.

The same procedure must be expanded slightly for the second grid. Copies of the spreadsheets which find answer sets follow.

	A	B	C	D	E
1	delta t =	0.2			epsilon =
2	rand start =	0.422654			0.1
3	start cell =	0.9999219			
4					
5	x =	-0.9995173	-0.9963444	0.9979226	
6		0.9999219	0	-0.9963022	
7					
8	g(x) =	0	0	1	
9		1		0	
10					
11	set sums R =	1	1	1	
12		1		0	
13					
14	set sums C =	1	0	1	
15		1		1	
16					
17	differences R	1	1	0	
18		0		0	
19					
20	differences C	1	0	0	
21		0		1	
22					
23	g(diff) =	1	0	0	
24		0		0	
25					
26	next x =	-0.9996138	-0.9970755	0.9983381	
27		0.9999375		-0.9970418	
28					
29	loop =	-0.9996138	-0.9970755	0.9983381	
30		0.9999375		-0.9970418	

	A	B	C	D	E
1	rand =	0.422654			
2	start =	0			
3	-1.0004641	-0.9888356	-0.9872757	0.988187	-0.98566
4	-1.0022425	-0.991578	0.9793051	x	x
5	-0.9911399	x	x	x	x
6	-0.9916013	0.985249	-0.991737	x	x
7	-0.9710076	x	0.9528802	x	x
8	0.987234	x	-1.0342069	0.9171555	-0.9171555
9	epsilon =	0.1			
10	delta t =	0.1			
11	g =				
12	0	0	0	1	0
13	0	0	1	x	x
14	0	x	x	x	x
15	0	1	0	x	x
16	0	x	1	x	x
17	1	x	0	1	0
18	R sums - i =				
19	1	1	1	0	1
20	1	1	0	x	x
21	0	x	x	x	x
22	1	0	1	x	x
23	0	x	0	x	x
24	0	x	1	0	1
25	C sums - i =				
26	1	0	1	0	0
27	1	0	0	x	x
28	1	x	x	x	x
29	1	0	1	x	x
30	1	x	0	x	x
31	0	x	1	0	0
32	next x =				
33	-1.0004177	-0.989952	-0.9885481	0.9893683	-0.987094
34	-1.0020182	-0.9924202	0.9813746	x	x
35	-0.9920259	x	x	x	x
36	-0.9924412	0.9867241	-0.9925633	x	x
37	-0.9739068	x	0.9575922	x	x
38	0.9885106	x	-1.0307862	0.9254399	-0.9254399
39	loop =				
40	-1.0004177	-0.989952	-0.9885481	0.9893683	-0.987094
41	-1.0020182	-0.9924202	0.9813746	x	x
42	-0.9920259	x	x	x	x
43	-0.9924412	0.9867241	-0.9925633	x	x
44	-0.9739068	x	0.9575922	x	x
45	0.9885106	x	-1.0307862	0.9254399	-0.9254399

Chapter 2--Hypergraphs and Neural Networks

2.1 Multiproducts in Neural Network Models

It is certainly desirable to have mathematically rigorous knowledge of the attractors of neural network models. In fact, for those models to be used in content addressable memory (static memories) or robot control, one usually seeks some assurance that the only attractors are constant trajectories built into the model and in particular that no limit cycle trajectories or chaotic trajectories occur. The goal of this chapter is to qualitatively describe in terms of hypergraphs certain models with the following property: all nonconstant trajectories asymptotically approach constant trajectories. The results contained here appeared in [JvdD].

Specifically, we shall study a version of the basic model of the previous Chapter,

$$dx_i/dt = -k_i x_i + p_i(g(x)) \tag{1.1}$$

namely,

$$dx_i/dt = k_i(I_i - x_i) + \sum_j a_{ij_1 j_2 \cdots j_p} g_{j_1} g_{j_2} \cdots g_{j_p} \tag{2.1}$$

In (2.1) we have abbreviated $g_j(x_j)$ by g_j; the summation is over certain permutations $\mathbf{j}$ of subsets of the indices $\{1, 2, \ldots, n\}$. Each I_i is a nonnegative constant. We call the expansion of g products and the constant $k_i I_i$ the *multiproduct* form of a high order model. Thus each $p_i(g)$ in (1.1) is partitioned in (2.1) into a constant component and a sum of products of components of g. The $p+1 \geq 2$ indices $\{i, j_1, j_2, \ldots, j_p\}$ in each permutation $\mathbf{j}$ are assumed distinct with $j_1 < j_2 < \ldots < j_p$ if $p > 1$. In each summand, a__ (the abbreviation of the full coefficient in (2.1)) is a constant coefficient with an associated *coefficient index set* $(i, j_1, j_2, \ldots, j_p)$. We think of the coefficient index set as having a

33

sign, + or -, namely the sign of a__ . Given a coefficient index set, we call the unordered set $\{i, j_1, j_2,..., j_p\}$ the associated *coefficient simplex*.

Clearly a set of systems of the form (2.1) with analogous coefficients $\{a__\}$ of the same sign comprises an equivalence class; we call such a set a *sign equivalence class*. We let N denote the number of coefficient simplexes in a sign equivalence class.

An example of (2.1) with n = 5 and N = 4 is

$$dx_1/dt = 1 - x_1 + a_{123}g_2g_3 + a_{134}g_3g_4 + a_{14}g_4 \tag{2.2}$$
$$dx_2/dt = 1 - x_2 + a_{213}g_1g_3$$
$$dx_3/dt = 1 - x_3 + a_{312}g_1g_2 + a_{314}g_1g_4$$
$$dx_4/dt = 1 - x_4 + a_{413}g_1g_3 + a_{41}g_1 + a_{45}g_5$$
$$dx_5/dt = 1 - x_5 + a_{54}g_4$$

The coefficient simplexes of (2.2) are $\{1, 2, 3\}$, $\{1, 3, 4\}$, $\{1, 4\}$, and $\{4, 5\}$.

Associated with every sign equivalence class of the form (2.2) is a *hypergraph* **H** with the following types of components. The hypergraph *vertices* are points labelled $\{v_1, v_2, ... , v_n\}$. The *barycenters* $\{b_1, b_2, ... , b_N\}$ of **H** are additional vertices, each thought of as corresponding to a coefficient simplex $\{i, j_1, j_2, ... , j_p\}$. Graphically each barycenter (open circle) is joined to these p+1 vertices by p+1 *edges* (line segments). Such an edge is signed as $(i, j_1, ... , j_p)$, that is, the sign of the corresponding a__ . A barycenter with all "+" edges or all "-" edges is said to be a *same sign barycenter*. Let us assume in (2.2) that a_{123}, a_{213}, a_{312}, a_{14}, a_{41}, a_{45}, and a_{54} are positive while a_{134}, a_{314}, and a_{413} are negative. Thus the four barycenters for (2.2) are same sign. The resulting hypergraph for (2.2) is shown in Fig. 2.1.

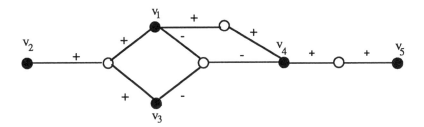

Fig. 2.1. The hypergraph associated with the dynamical system (2.2) with $a_{123}, a_{213}, a_{312}, a_{14}, a_{41}, a_{45}$, and a_{54} positive; and a_{134}, a_{314}, and a_{413} negative.

As stated above, the purpose of this chapter is to establish a result of the following type: if every barycenter of **H** is same sign and if certain "balanced cycle" conditions are met, then we can guarantee that any nonconstant trajectory for (2.2) must asymptotically approach some constant trajectory. In particular, cyclic and chaotic trajectories are precluded.

2.2 Paths, Cycles, and Volterra Multipliers

Suppose in a hypergraph **H** constructed as above that there exist $q+1 \geq 2$ distinct vertices $\{v_1, \dots, v_{q+1}\}$; q distinct barycenters $\{b_1, \dots, b_q\}$ and 2q distinct signed edges $\{e_1, e_2, \dots, e_{2q}\}$ with e_1 connecting b_1 to v_1, e_2 connecting b_1 to v_2, e_3 connecting b_2 to v_2, and so on up to e_{2q} connecting b_q to v_{q+1}. We call the triple set

$$[\{v_1, \dots, v_{q+1}\}, \{b_1, \dots, b_q\}, \{e_2, e_4, \dots, e_{2q}\}]$$

a *vertex-to-vertex (q+1)-path, or path, from* v_1 *to* v_{q+1}. Of course, the components

$$[\{v_{q+1}, \dots, v_1\}, \{b_q, \dots, b_1\}, \{e_{2q-1}, \dots, e_3, e_1\}]$$

constitute a second (q+1)-path from v_{q+1} to v_1. A hypergraph is *connected* if any two vertices are in some path. If the hypergraph associated with (2.1) is not connected, then (2.1) is actually more than one dynamical system; so we deal only with systems with connected hypergraphs. A (q+1)-*cycle*, q+1 ≥ 2, or simply *cycle*, is defined just like a (q+2)-path except that $v_{q+2} = v_1$. To every cycle corresponds a reverse cycle; if every barycenter is same sign, then the number of negative edges in a cycle equals the number of negative edges in its reverse cycle. Given a cycle, we call the absolute value of the product of the a__ coefficients corresponding to its edges the *weight* of the cycle. If a cycle and its reverse have equal weights, the cycle is *balanced*. This holds in (2.2) if $|a_{123}a_{314}| = |a_{134}a_{312}|$ and $|a_{14}a_{413}| = |a_{41}a_{134}|$ Note the relative positions of edges in Fig. 2.1 corresponding to the entries in these equations.

Suppose all barycenters for (2.1) have the same sign property. If there are positive constants $\{\lambda_1, \lambda_2, \ldots, \lambda_n\}$ for each barycenter which satisfy $\lambda_i |a_{i\_\_}| = \lambda_j |a_{j\_\_}|$ (i, j are any indices in a coefficient simplex), then the constants $\{\lambda_i\}$ are called *Volterra multipliers* (for an account of related linear Volterra multipliers, see [R] and the numerous precedent papers by the same author on this topic). In the case of (2.2), we require

$$\lambda_1 a_{123} = \lambda_2 a_{213} = \lambda_3 a_{312}$$
$$\lambda_1 a_{134} = \lambda_3 a_{314} = \lambda_4 a_{413} \qquad\qquad (2.3)$$
$$\lambda_1 a_{14} = \lambda_4 a_{41}$$
$$\lambda_4 a_{45} = \lambda_5 a_{54}$$

It follows that Volterra multipliers exist satisfying (4) precisely if $|a_{123}a_{314}|$ $= |a_{134}a_{312}|$ and $|a_{14}a_{413}| = |a_{41}a_{134}|$. In fact, if every barycenter in the hypergraph of (2.1) is same sign, then Volterra multipliers exist for (2.1) if and only if every cycle in (2.1) is balanced; this will be established first for some special cases.

LEMMA 2.1. (Acyclic nonlinear case) Suppose the hypergraph of (2.1) has no cycles and every barycenter in the hypergraph is same sign. Then the system admits Volterra mulitpliers.

PROOF. Let λ_1 be any positive number. Suppose there is a barycenter connected to both v_1 and some v_j. Define $\{\lambda_j\}$ by the equation

$\lambda_1 a_{1j-} = \lambda_j a_{j1-}$. Due to the same sign condition, each $\lambda_j > 0$. Clearly this process can be propagated through the acyclic hypergraph to specify all $\{\lambda_i\}$.

QED

LEMMA 2.2. (Linear case) Suppose every barycenter in the hypergraph of (2.1) is same sign and connected to exactly two vertices. Then Volterra multipliers exist for (2.1) iff every cycle is balanced.

PROOF. The case with no cycle is covered in Lemma 1. So we assume that there is a cycle with vertices $\{v_1,...v_j,...v_{j+k}\}$.

Suppose Volterra multipliers exist. On the path from v_1 to v_j (with vertices $v_1, v_2,..., v_{j-1}, v_j$) we have multipliers λ_1 through λ_j satisfying

$$\lambda_i a_{i\,i+1} = \lambda_{i+1} a_{i+1\,i} \tag{2.4}$$

so

$$\lambda_j = [(a_{12}a_{23}...a_{j-1\,j})/(a_{j\,j-1}...a_{32}a_{21})]\,\lambda_1 \tag{2.5}$$

Since every barycenter is same sign, each λ_i defined along the path is a positive multiple of λ_1. For the path from v_1 to v_j with vertices $v_1, v_{j+k}, ..., v_{j+1}, v_j$, the analogy of (2.4) implies

$$\lambda_j = [(a_{1\,j+k}...a_{j+1\,j})/(a_{j+k\,1}...a_{j\,j+1})]\,\lambda_1 \tag{2.6}$$

Thus the balanced cycle condition is a consequence of (2.5) and (2.6).

For the converse, assume the balanced cycle condition, choose $\lambda_1 > 0$, and define other λ_i using (2.4) along various paths. The same sign condition implies each $\lambda_i > 0$ and the balanced cycle condition implies the definitions are independent of path choice.

QED

THEOREM 2.1. Suppose every barycenter in the hypergraph $\mathbf{H}$ of (2.1) is same sign. Then Volterra multipliers exist for (2.1) iff every cycle in (2.1) is balanced.

PROOF. Suppose Volterra multipliers exist. We proceed as in Lemma 2.2, except that the associated coefficient simplexes might be of higher cardinality. Choose a cycle with vertices $\{v_1,...v_j,...v_{j+k}\}$. The analog of (2.4) is

$$\lambda_i a_{i-\_} = \lambda_{i+1} a_{i+1-\_} \tag{2.7}$$

For the converse, assume all cycles in **H** are balanced. As in the proof of Lemma 2.2, any λ_j can be defined as a positive multiple of λ_1 by calculations along a path connecting vertices v_1 and v_j. That the definition of λ_j is independent of path follows from analogues of (2.5) and (2.6) with higher order indices.   QED

2.3 The Cohen-Grossberg Function Γ

We shall use a version of the Cohen and Grossberg [CG] function Γ for (2.1) with Volterra multipliers $\{\lambda_i\}$:

$$\Gamma = -\sum_{i=1}^{n} \lambda_i \int_{-\varepsilon_i}^{x_i} k_i (I_i - y) g_i'(y) \, dy - \sum_{\text{barycenters}}^{N} \lambda_i a_{ij_1 j_2..j_p} \, g \, g_{j_1} g_{j_2} \cdots g_{j_p} \tag{2.8}$$

Here g' denotes the derivative of g except $g_i'(\pm\varepsilon_i) \equiv 0$. In each barycenter summand we use any one of the (equal) coefficients $\lambda_i a_{i-\_}$. Thus for the special case (2.2) with balanced cycles, Γ in (2.8) becomes

$$\Gamma = -\sum_{i=1}^{4} \lambda_i \int_{-\varepsilon_i}^{x_i} k_i (I_i - y) g_i'(y) \, dy - \lambda_1 a_{123} g_1 g_2 g_3 - \lambda_1 a_{134} g_1 g_3 g_4 - \lambda_1 a_{14} g_1 g_4 - \lambda_4 a_{45} g_4 g_5$$

We note that Γ is continuous and that Γ is differentiable along an arbitrary trajectory for (2.2) except when some i component of that trajectory passes through $\pm\varepsilon_i$. Except for such points we have

$$d\Gamma/dt = -\sum_{i=1}^{n} \lambda_i g_i'(x_i) \dot{x}_i^2 \tag{2.9}$$

Considering the behavior of Γ (2.9) and Theorems 1.1, 1.2, and 2.1, we are ready to prove

THEOREM 2.2. If the hypergraph of (2.1) has same sign barycenters and balanced cycles (and therefore Volterra multipliers in (2.7)), then every trajectory for (2.1) must be or must asymptotically approach a constant trajectory.

PROOF. The application of Theorem 1.1 to such a system guarantees the existence of a finite attractor region. Theorem 2.1 guarantees the existence of Volterra multipliers $\{\lambda_i\}$ and leads to the specification of Γ (2.8).

Suppose $x(t)$ is a nonconstant <u>cyclic</u> trajectory. From (2.9) we see $d\Gamma/dt = 0$ along $x(t)$ except at points on the boundary of T_ε. Also, $x(t)$ cannot lie entirely in one orthant relative to T_ε (where dynamics are linear and not cyclic). Let $x(t_0) \in T_\varepsilon$ with $|x_i(t_0)| < \varepsilon_i$ for at least one index i. If $|x_i(t_0)| < \varepsilon_i$ for all i, then (2.9) implies $x(t)$ must actually be a constant trajectory (since all $g_i' > 0$ at $x(t_0)$). Thus there must be nonempty index sets I and J such that i $\in I$ implies $|x_i(t_0)| < \varepsilon_i$ while $j \in J$ implies $|x_j(t_0)| \geq \varepsilon_j$. Furthermore, $g_i' = 1/(2\varepsilon_i)$ implies each such x_i must actually be constant for time t near t_0. But then $dx_j/dt = k_j(I_j - x_j) + d_j$ $(d_j = $ a constant) for t near t_0. Not all such x_j can be constant, so every such $|x_j(t)|$ must asymptotically approach a constant $(= I_j + d_j/k_j)$ or some $|x_j(t)|$ must decrease until $|x_j(t)| < \varepsilon_j$. The first cannot occur for a cyclic trajectory; the second yields a time when g_j' is negative (contradicting $d\Gamma/dt = 0$). Thus (2.1) with Volterra multipliers cannot admit a cyclic trajectory.

Of course, Γ (a continuous function) evaluated over the (finite diameter) attractor region of (2.1) is bounded. Thus, along <u>any nonconstant</u> trajectory $x(t)$, $d\Gamma/dt$ must asymptotically approach 0. Select an infinite countable sequence of times $t_1, t_2, ...$ with $t_{j+1} - t_j > 1$ such that

$$\lim_{j \to \infty} d\Gamma/dt_{x(t_j)} = 0$$

Since all the points $x(t_j)$ are in the attractor region (a compact set), we can select an infinite subsequence of points that converge to a point x_∞ in the attractor region. Since trajectories for (2.1) depend continuously on initial conditions, the trajectory $x_\infty(t)$ starting at x_∞ must have $d\Gamma/dt = 0$.

If $x_\infty(t)$ does not lie in an orthant relative to T_ϵ for all future time, then an argument using $d\Gamma/dt = 0$ along $x_\infty(t)$ and otherwise completely analogous to the above cyclic trajectory case yields a contradiction. If $x_\infty(t)$ does lie in an orthant relative to T_ϵ for all future time, then Theorem 1.2 guarantees that $x_\infty(t)$ must asymptotically approach a constant trajectory.   QED

2.4 The Foundation Function Φ

Suppose a system of the form (2.1) has Volterra multipliers $\{\lambda_i\}$. We can define a *foundation function* Φ by

$$\Phi = \sum_{j \in \text{ coefficient simplexes}} \kappa_j g_{j_1} g_{j_2} \cdots g_{j_p} \tag{2.10}$$

where κ_j are constants chosen to satisfy $p_i(g) = -\lambda_i^{-1}\partial\Phi/\partial g_i$. This choice is possible, that is, consistent for all terms representing a given coefficient simplex, precisely due to the existence of the Volterra multipliers.

As a generalization, we next study the model

$$dx_i/dt = -k_i x_i + p_i(g(x)) \tag{1.1}$$

with the restriction that the system functions $p_i(g)$ arise as multiples of first-order partial derivatives of a foundation function.

THEOREM 2.3. Suppose we are given a neural network model of the form (1.1) with each $p_i(g) = -\lambda_i^{-1}\partial\Phi/\partial g_i$ for n positive constants $\{\lambda_i\}$ and a continuously differentiable function Φ. Then every trajectory for the model must be a constant trajectory or must asymptotically approach a constant trajectory.

PROOF. The generalization of (2.8) which we need is

$$\Gamma = -\sum_{i=1}^{n} \lambda_i \int_{-\varepsilon_i}^{x_i} k_i(I_i - y)g_i'(y)\,dy \; + \; \Phi(g(x)) \qquad (2.11)$$

Along an arbitrary trajectory for (1.1) the derivative of Γ in (2.11) is still given by (2.9) except on the boundary of T_ε. The dynamics of Γ here are completely parallel with the dynamics of Γ as developed in the proof of Theorem 2.2. QED

An important special case of models included in Theorem 2.3 are set selection models of the previous chapter:

$$dx_i/dt = -x_i + 1 - (2/n_i) \sum_{\substack{S_j \\ i \in S_j}} \; \sum_{\substack{k \in S_j \\ k \neq i}} g_k(x_k) \qquad (1.5)$$

THEOREM 2.4. Suppose we are given a set selection model of the form (1.5). Every trajectory for the model must be a constant trajectory or must asymptotically approach a constant trajectory.

PROOF. To fulfill the conditions of Theorem 2.3 we need only define

$$\Phi = \sum_{S_j} \; \sum_{\substack{k,l \in S_j \\ k \neq l}} g_k(x_k)g_l(x_l)$$

and $\lambda_i = -n_i/2$. \hfill QED

A foundation function Φ is like a physical potential in that its gradients determine system dynamics. If Φ is regarded as some generalization of (2.10) or perhaps a multinomial in g of the form

$$\Phi = \sum_{j=1}^{N} \Phi_j(g)$$

then each $\Phi_j(g)$ corresponds to some sort of interaction among a subset of neurons. If each foundation function summand $\Phi_j(g)$ is a product of positive integer powers of certain $\{g_i\}$, and if the neurons represented in $\Phi_j(g)$ are all simultaneously saturated (all such $g_i = 1$), then each state x_i is driven in part by that mutually competitive or mutually cooperative interaction, concepts used elsewhere in the neural network literature.

The Cohen-Grossberg function for (2.1) has been exploited by Hirsch [Hi] using pde sign conditions. Our conditions, being couched instead in terms of hypergraphs and cycles, offer an alternative suggestion of how neural networks might be organized. The hypergraph conditions are not equivalent to the pde conditions. For example, if $|a_{123}a_{314}| = |a_{134}a_{312}|$ and $|a_{14}a_{413}| = |a_{41}a_{134}|$ in (2.2), then Theorem 2.2 can be applied. But we see in (2.2) that the terms $\partial(dx_1/dt)/\partial x_3$ and $\partial(dx_3/dt)/\partial x_1$ are not of constant sign for all x, thus violating a pde sign condition on p. 119 of [Hi].

It is important to recall that the constant trajectories in Theorem 2.2 might include stable constant trajectories inside T_ϵ. Also, it is not difficult to construct simple versions of (2.1) with limit cycles and, of course, no foundation function. For examples, see accounts of systems (1.3), (1.6), and (1.7) in the previous chapter.

In summary, for some applications in content addressable memory and robot arm control, neural networks should only have constant trajectories as attractors. As brought out by system (2.1) and Theorems 2.1 and 2.2, one design route to this feature is the organization of neurons in mutually inhibitory or mutually excitatory interaction subsets (hypergraphs with same sign barycenters) together with the balanced cycle condition. More generally, Theorem 2.3 states that (2.10) has only constant trajectories as attractors if a foundation function Φ and positive constants $\{\lambda_i\}$ (Volterra multipliers) exist so each $p_i(g) = -\lambda_i^{-1}\partial\Phi/\partial g_i$.

2.5 The Image Product Formulation of High Order Neural Networks

Our formulation of high order neural networks has dealt primarily with the multiproduct form

$$dx_i/dt = k_i(I_i - x_i) + \sum_j a_{ij_1j_2 \ldots j_p} g_{j_1} g_{j_2} \cdots g_{j_p} \qquad (2.1)$$

The goal of this section is to derive an algebraic equivalence between this form and a sum of products, n terms in each product, each multiplicand in each product being either g_i or $1-g_i$; in certain cases the latter form turns out to be more tractable in the analysis of neural models.

Given 2^n real numbers a__, let us form the algebraic sum as a multinomial in the components of the n-vector $g = (g_1, g_2,\ldots, g_n)$

$$A_a(g) = a_0 + a_1g_1 + a_2g_2 + \ldots + a_ng_n + a_{12}g_1g_2 + \ldots + a_{n-1n}g_{n-1}g_n + \ldots$$
$$+ a_{12\ldots n}g_1g_2\cdots g_n \qquad (2.12)$$

Alternatively, given any 2^n real numbers b__ indexed by the binomial n-strings, we can form the sum

$$B_a(g) = b_{00\ldots0}(1-g_1)(1-g_2)\ldots(1-g_n) + b_{10\ldots0}g_1(1-g_2)\ldots(1-g_n) + \ldots + b_{11\ldots1}g_1g_2\cdots g_n$$

$$(2.13)$$

We refer to such a sum of products (2.13) of g components as an *image product*.

Finally, we can suppose that a__ and b__ are related as in the following table.

$$b_{00...0} = a_0 \tag{2.14}$$
$$b_{10...0} = a_0 + a_1$$
$$b_{01...0} = a_0 + a_2$$

...

$$b_{00...1} = a_0 + a_n$$
$$b_{110...0} = a_0 + a_1 + a_2 + a_{12}$$
$$b_{101...0} = a_0 + a_1 + a_3 + a_{13}$$

...

$$b_{100...1} = a_0 + a_1 + a_n + a_{1n}$$
$$b_{1110...0} = a_0 + a_1 + a_2 + a_3 + a_{12} + a_{13} + a_{23} + a_{123}$$

...

$$b_{11...1} = a_0 + a_1 + a_2 +...+ a_n + a_{12} + a_{13} +...+ a_{n-1\,n} + a_{123} + ... + a_{12...n}$$

With this algebraic machinery in place we are in a position to simplify an analysis of the conventional formulation of the high order neural network model. It can be proved that if $\underline{b}$ is derived from $\underline{a}$ as in (2.14), then $A_a(g) = B_b(g)$.

As an example, let us rewrite the two-dimensional neural network model

$$dx_1/dt = -x_1 + g_1(1-g_2) - (1-g_1)g_2$$
$$dx_2/dt = -x_2 - g_1(1-g_2) + (1-g_1)g_2$$

in the A format

$$dx_1/dt = -x_1 + g_1 - g_2$$
$$dx_2/dt = -x_2 - g_1 + g_2$$

Thus in dx_1/dt, $a_0 = 0$, $a_1 = 1$, $a_2 = -1$, and $a_{12} = 0$ while $b_{00} = 0$, $b_{10} = 1$, $b_{01} = -1$, and $b_{11} = 0$. In each of the four orthants relative to T_ε we have $dx_i/dt + x_i = 0$ or ± 1. Which value occurs is easier to see in the B format (product form). This improvement is especially clear in higher dimensional models. For example, outside the transition zone T_ε in 10-dimensional space, the term

$$(1-g_1)g_2g_3g_4(1-g_5)(1-g_6)g_7(1-g_8)g_9g_{10}$$

in a model is clearly 0 except in the (-,+,+,+,-,-,+,-,+,+) orthant relative to T_ε ; in that orthant the term is 1. The A form

$$g_2g_3g_4g_7g_9g_{10} - g_1 g_2g_3g_4g_7g_9g_{10} +\cdots+ g_1g_2g_3g_4g_5g_6g_7g_8g_9g_{10}$$

of the same term would not be so transparent.

　　　We establish the equivalence of the two forms in a theorem.

THEOREM 2.5. If $\underline{b}$ is derived from $\underline{a}$ as in (2.14), then $A_a(g) = B_b(g)$.

PROOF.　Clearly if each $g_i = 0$ or 1, then $A_a(g) = B_b(g)$. Our task is to derive the algebraic identity $A_a(g) = B_b(g)$, valid for all n-vectors g.

　　　Given (2.12) we may write $A_a(g) = a* \cdot g$ where * denotes transpose and

$$g* = (1, g_1, g_2,..., g_n, g_1g_2, g_1g_3,..., g_1g_2\cdots g_n)$$

Let a second variable vector h be defined by

$$h* = ((1-g_1)(1-g_2)...(1-g_n), g_1(1-g_2)...(1-g_n),..., g_1g_2\cdots g_n)$$

Thus (2.13) is equivalent to $B_b(g) = b* \cdot h$. Our proof reduces to showing $a* \cdot g = b* \cdot h$ regardless of the magnitudes or signs of the components of g.

　　　We define a $2^n \times 2^n$ matrix M each entry of which is 0 or 1 by $\underline{b} = M\underline{a}$.

　　　Note that the 2^n vectors of the form g generated by selecting various g with 0,1 components are linearly independent. Likewise the 2^n vectors of the form h generated by selecting various g with 0,1 components are linearly independent. Let a second $2^n \times 2^n$ matrix N be defined to satisfy h = Ng for all such selections.

　　　If every component of g is 0 or 1, then the fact that $A_a(g) = B_b(g)$ is equivalent to

$$A_a(g) = a* \cdot g = B_b(g) = b* \cdot h = b*Ng = a*M*Ng \qquad (2.15)$$

Since (2.15) must hold for all g generated from g with 0,1 components and all a, it follows that M*N must actually be the identity matrix. This implies that $A_a(g) = B_b(g)$ for all g, regardless of the component sign or magnitude.   QED

The implication of Theorem 2.5 is that if the $p_i(g)$ part of a neural network model (1.1) is of the form (2.12), then we can replace that sum with (2.13). (Theorem 2.5 actually applies to more general $p_i(g)$ than in (2.1) because in the proof of Theorem 2.5 we have not used the property i $\notin$ $\{j_1, j_2, ..., j_p\}$.) The advantage of the B formulation (2.13) is that at most one summand is on (= 1) in each orthant relative to T_ϵ. This, in turn, facilitates the design of models with chosen constant attractors and cyclic attractors.

We conclude with some tables of cases of (2.14). The explicit form of (2.14) for n = 2 is

$b_{00} = a_0$
$b_{10} = a_0 + a_1$
$b_{01} = a_0 + a_2$
$b_{11} = a_0 + a_1 + a_2 + a_{12}$

The explicit form of (2.14) for n = 3 is

$b_{000} = a_0$
$b_{100} = a_0 + a_1$
$b_{010} = a_0 + a_2$
$b_{001} = a_0 + a_3$
$b_{110} = a_0 + a_1 + a_2 + a_{12}$
$b_{101} = a_0 + a_1 + a_3 + a_{13}$
$b_{011} = a_0 + a_2 + a_3 + a_{23}$
$b_{111} = a_0 + a_1 + a_2 + a_3 + a_{12} + a_{13} + a_{23} + a_{123}$

The explicit form of (2.14) for n = 4 is

$b_{0000} = a_0$
$b_{1000} = a_0 + a_1$
$b_{0100} = a_0 + a_2$
$b_{0010} = a_0 + a_3$
$b_{0001} = a_0 + a_4$
$b_{1100} = a_0 + a_1 + a_2 + a_{12}$
$b_{1010} = a_0 + a_1 + a_3 + a_{13}$
$b_{1001} = a_0 + a_1 + a_4 + a_{14}$
$b_{0110} = a_0 + a_2 + a_3 + a_{23}$
$b_{0101} = a_0 + a_2 + a_4 + a_{24}$
$b_{0011} = a_0 + a_3 + a_4 + a_{34}$
$b_{1110} = a_0 + a_1 + a_2 + a_3 + a_{12} + a_{13} + a_{23} + a_{123}$
$b_{1101} = a_0 + a_1 + a_2 + a_4 + a_{12} + a_{14} + a_{24} + a_{124}$
$b_{1011} = a_0 + a_1 + a_3 + a_4 + a_{13} + a_{14} + a_{34} + a_{134}$
$b_{0111} = a_0 + a_2 + a_3 + a_4 + a_{23} + a_{24} + a_{34} + a_{234}$
$b_{1111} = a_0 + a_1 + a_2 + a_3 + a_4 + a_{12} + a_{13} + a_{14} + a_{23} + a_{24} + a_{34} + a_{1234}$

PROBLEMS

1. Consider the model

$$dx_1/dt = .5 - x_1 + g_3 - g_2g_3 - g_3g_4$$
$$dx_2/dt = .5 - x_2 - g_1g_3$$
$$dx_3/dt = .5 - x_3 + g_1 - g_1g_2 - g_1g_4$$
$$dx_4/dt = .5 - x_4 - g_1g_3$$

Sketch the associated hypergraph. Does every nonconstant trajectory asymptotically approach a constant trajectory? Are all constant trajectories that lie in orthants relative to T_ε necessarily stable? Assume $\varepsilon = .1$. Find the constant trajectories in orthants relative to T_ε.

2. Consider the model

$$dx_1/dt = -.5 - x_1 + g_2 - g_2g_3 + g_2g_3g_4$$
$$dx_2/dt = -1 - x_2 + 2g_1 - 2g_1g_3 + 2g_1g_3g_4$$
$$dx_3/dt = +1 - x_3 - 3g_1g_2 + 3g_1g_2g_4$$
$$dx_4/dt = -1 - x_4 + 6g_1g_2g_3$$

Sketch the associated hypergraph. Does every nonconstant trajectory asymptotically approach a constant trajectory? Are all constant trajectories that lie in orthants relative to T_ε necessarily stable? Assume $\varepsilon = .1$. Find the constant trajectories in orthants relative to T_ε.

3. Analyze

$dx_1/dt = +1 -x_1 - [g_2+g_3+g_4 +g_5+g_9]$

$dx_2/dt = +1 -x_2 - [g_1+g_3+g_4 +g_7+g_{10}]$

$dx_3/dt = +1 -x_3 - [g_1+g_2+g_4 +g_6]$

$dx_4/dt = +1 -x_4 - [g_1+g_2+g_3 +g_8]$

$dx_5/dt = +1 -x_5 - [g_1+g_6+g_9]$

$dx_6/dt = +1 -x_6 - [g_3]$

$dx_7/dt = +1 -x_7 - [g_2+g_8+g_{10}]$

$dx_8/dt = +1 -x_8 - [g_4+g_7]$

$dx_9/dt = +1 -x_9 - [g_1+g_5+g_{10}]$

$dx_{10}/dt = +1 -x_{10} - [g_2+g_7+g_9]$

ANSWERS

1. The associated hypergraph is

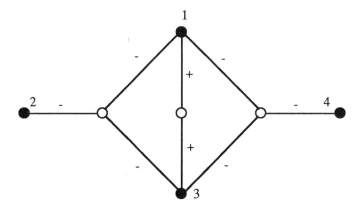

Because the hypergraph has same sign barycenters and balanced cycles, every nonconstant trajectory must asymptotically approach a constant trajectory. Around every constant trajectory outside the transistion zone T_ε, the system is linear with approximation matrix -I; stability follows.

At a constant trajectory,

$x_1 = .5 + g_3 - g_2g_3 - g_3g_4$
$x_2 = .5 - g_1g_3$
$x_3 = .5 + g_1 - g_1g_2 - g_1g_4$
$x_4 = .5 - g_1g_3$

Assuming $g_1 = 0$: $\Rightarrow x_2 = .5$, $x_4 = .5 \Rightarrow g_3 = 1$, $x_3 = .5 \Rightarrow x_1 = -.5$, $g_1 = 0 \Rightarrow x_1 = -.5$, the only solution with $g_1 = 0$.

Assuming $g_3 = 0$: $\Rightarrow x_2 = .5$, $x_4 = .5 \Rightarrow g_1 = 1$, $x_1 = .5 \Rightarrow x_3 = -.5$, $g_3 = 0 \Rightarrow x_1 = .5$, the only solution with $g_3 = 0$.

Assuming $g_1g_3 = 1$: $\Rightarrow x_2 = -.5$, $x_4 = -.5 \Rightarrow g_2 = g_4 = 0 \Rightarrow g_1 = 1$, $x_1 = 1.5$, $g_3 =$

1, $x_3 = 1.5$, the only solution with $g_1g_3 = 1$.

2. The associated hypergraph is

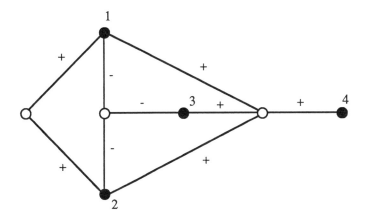

Because the hypergraph has same sign barycenters and balanced cycles, every nonconstant trajectory must asymptotically approach a constant trajectory. Around every constant trajectory outside the transistion zone T_ε, the system is linear with approximation matrix -I; stability follows.

At a constant trajectory,

$x_1 = -.5 + g_2 - g_2g_3 + g_2g_3g_4$
$x_2 = -1 + 2g_1 - 2g_1g_3 + 2g_1g_3g_4$
$x_3 = +1 - 3g_1g_2 + 3g_1g_2g_4$
$x_4 = -1 + 6g_1g_2g_3$

Assuming $g_1 = 0$: $\Rightarrow x_3 = 1$, $x_4 = -1 \Rightarrow g_3 = 1$, $g_4 = 0 \Rightarrow x_2 = -1$, $g_2 = 0 \Rightarrow x_1 = -.5$, the only solution with $g_1 = 0$.

Assuming $g_2 = 0$: $\Rightarrow x_3 = 1$, $x_4 = -1 \Rightarrow g_3 = 1$, $g_4 = 0 \Rightarrow x_1 = -.5$, $g_1 = 0 \Rightarrow x_2 = -1$, same solution.

Assuming $g_3 = 0$: $\Rightarrow x_4 = -1$, $g_4 = 0 \Rightarrow x_3 = -2$, $g_1g_2 = 1 \Rightarrow x_2 = 1$, $g_2 = 1 \Rightarrow x_1 = .5$, $g_1 = 1$, the only solution with $g_3 = 0$.

Assuming $g_1 g_2 g_3 = 1: \Rightarrow x_4 = 5, g_4 = 1 \Rightarrow g_4 = 1 \Rightarrow x_1 = .5, x_2 = 1, x_3 = 3$, the only solution with $g_1 g_2 g_3 = 1$.

3. This model is associated with a set selection problem, namely: consider the array of elements

```
1   2   3   4
5       6
    7       8
9   10
```

and choose exactly one element from each row and each column. If all $\{\lambda_i\}$ are taken to be 1, the foundation function is

$$\Phi = g_1 g_2 + g_1 g_3 + g_1 g_4 + g_1 g_5 + g_1 g_9 + g_2 g_3 + g_2 g_4 + g_2 g_7 + g_2 g_{10} + g_3 g_4 + g_3 g_6 + g_4 g_8$$
$$+ g_5 g_6 + g_5 g_9 + g_7 g_8 + g_7 g_{10} + g_9 g_{10}.$$

Thus all trajectories asymptotically approach constant trajectories. An example of a constant trajectory representing an answer set is $x_4 = x_6 = x_7 = x_9 = 1$, all other $x_i = -1$.

Chapter 3--The Memory Model

3.1 Dense Memory with High Order Neural Networks

In this chapter we develop the *memory model*, a dynamical system neural network model which simulates memory retrieval. The time evolution of a trajectory of such a system is mathematical *recognition*, meaning convergence to one of several attractors referred to as mathematical *memories*. The attractors are prespecified constant trajectories or limit cycles. Thus mathematical recognition amounts to convergence from an input vector in n-dimensional space to one of the memories represented as an n-vector with components ±1.

The *memory model* can store, using n neurons, any number M, $1 \leq M \leq 2^n$, of any of the binomial ±1 n-strings as constant attractor memories; in a schematic representation the model requires only $5n+M(1+2n)$ edges. The gain function used in the memory model has range $[0, 1]$ and can be a ramp function or a ramp function with jump (see Fig. 3.8). If the slope or *gain* of the gain function at 0 is sufficiently high, the only stable constant attractors are the memories. The memory model can be used in error-correcting decoding of any binary string code.

Statistical analysis [JP] of more than a billion tests of various versions of the memory model has shown that it is practically as accurate as Euclidean distance at converging from an initial state (corrupted code word) to the nearest of the given memories. In these tests the memory model has been used as an n-dimensional dynamical system decoder for standard binary codes with n = 7, 11, 15, 23, and 24. As explained in the next chapter, a standard code consists of a choice of some (typically $2^{.5n}$) of the 2^n binary n-strings. These tests involved the following steps.

1. Select at random one of the memories (a code word written as an n-string with ±1 entries).

2. Corrupt the code word with Gaussian noise (using various amplitudes

53

and distributions) or uniformly random noise with bounds.

  3. Use the corrupted code word as an initial state for the memory model and allow the model to iterate as a dynamical system until a neighborhood of a steady state is reached (typically 10 to 50 time steps, depending upon n and the number of memories). Record whether or not the steady state is the original memory.

  Of course, with sufficient noise the neural network or brute force Euclidean distance decoding will not converge to the original code word. However, the tests have shown that with a signal to noise ratio of 5 to 10 dB (explained in the next chapter), the memory model is 10 to 100 times less likely to make an error (converge to the wrong code word) than a conventional hard algebraic decoder for the same code. This seems partly due to the fact that a hard algebraic decoder must first round corrupted signals to binary form, then proceed to compute a series of products and sums of input entries in order to determine the intended code word. The initial rounding amounts to a loss of information.

  It should be emphasized that in certain cases the memory model will converge to a code word which is <u>not</u> the closest. An example is given in §4 of this chapter. This shows that the memory model is not Euclidean distance in disguise. However, computer experiments have statistically proven that with proper choice of system coefficients this type of error is very unlikely.

  If it were physically possible to implement the memory model as microelectronic or microoptical chip, then the model might have extensive commercial application in telecommunications. That is, a chip based on the memory model would be more accurate and faster than conventional code recognition algorithms carried out on a serial computer.

  A model of recognition is said to have *content addressable memory* because partial knowledge of a memory leads to the complete retrieval of the memory through system dynamics. Thus neural network recognition is formally equivalent to the general concept of error correction of binary code by means of an analog dynamical system, much in contrast to the use of serial algebraic algorithms in the conventional theory of error-correcting codes.

  All attractor trajectories in the memory model (constant trajectories or limit cycles) are specified in advance (the model is "hardwired" at the outset, not trained). Each constant attractor trajectory is one of the 2^n points in n-dimensional space with coordinates ± 1, that is, a vertex of the n-cube with vertex component values ± 1. Each cyclic attractor trajectory is associated with a cyclic tour of some orthants of n-space, each orthant labelled by a vertex of the n-cube. All attractors are disjoint in the sense that different attractors correspond to disjoint vertex sets.

As described below, a feedback loop is associated with each constant memory in a memory model. Learning a pattern might be realized as driving the device to a particular state and holding it there while the associated feedback loop is somehow electronically activated ("burned in").

3.2 High order neural network models

As specified in the previous chapters, the activity levels of n mathematical neurons are represented by a point x in n-dimensional state space, that is, the space consisting of n-tuples $x = (x_1, x_2, \ldots, x_n)$ of real numbers. Choose some positive $\varepsilon < 1$. Define a *gain function* $g_i = g_i(x_i)$ by $g_i = 0$ if $x_i \leq -\varepsilon$; $g_i = 1$ if $x_i \geq \varepsilon$; and g_i is continuously differentiable and increasing for $|x_i| < \varepsilon$. We call $g_i'(0)$ the *gain* of g_i. Suppose for each i = 1, 2, $\ldots$, n that p_i is a continuously differentiable function of n-space and k_i is a positive constant.

$$dx_i/dt = -k_i x_i + p_i(g(x)) \qquad \text{where } i = 1,\ldots,n \qquad (3.1)$$

This system has constant trajectories solving the nonlinear equations

$$x_i = k_i^{-1} p_i(g(x)) \qquad \text{where } i = 1,\ldots,n \qquad (3.2)$$

As already mentioned, (3.1) is actually 3^n dynamical systems in 3^n regions of state space partitioned by the 2^n hyperplanes $\{x \mid \text{some } x_i = \pm\varepsilon \}$. In each region, trajectories for (3.1) exist and are unique. Any vector-valued function of time which is made up of such trajectories is regarded as a trajectory for (3.1).

The *transition zone* for (3.1) is the open set T_ε of points x in n-space with at least one component x_i satisfying $|x_i| < \varepsilon$. (In two-dimensional space, T_ε is the "+" shaped region in Fig. 1.1.) Clearly outside T_ε each $g_i(x_i)$ is constant ($= 0$ or 1) and thus each $p_i(g(x))$ is constant. It is also clear that the complement of T_ε in n-space consists of 2^n (closed) components, each naturally a subset of an orthant of n-space; we refer to each such component as an *orthant relative to* T_ε. Trajectories for (3.1) in an orthant O relative to T_ε are simply pieces of trajectories of the linear (constant coefficient) dynamical system $dx_i/dt = k_i(-x_i+c_i)$ for some constant c_i; $c = (c_1, c_2, ...,c_n)$ itself might or might not lie in O. If c solves (3.2), then c lies in O and is a constant attractor trajectory. We assume throughout that no such c_i equals $\pm\varepsilon$ (otherwise decrease ε as needed; this would not change any c). This insures that trajectories entering T_ε from an orthant relative to T_ε do so transversally (not tangentially).

Note that any product $\mathbf{I}$ of the 2^n possible products of n numbers, the ith of which is g_i or $1-g_i$, is itself a number with $0 \le \mathbf{I} \le 1$. In exactly one orthant relative to T_ε, such a product $\mathbf{I}$ is 1, and in all other orthants relative to T_ε, $\mathbf{I}$ is 0. For example, in four-dimensional space the product $g_1(1-g_2)(1-g_3)g_4$ is *turned on* (= 1) in the (+,-,-,+)-orthant containing the point (1, -1, -1, 1) and *turned off* (= 0) in all other 15 orthants. As described below, including a particular n-string as a constant memory entails including its associated product in the system functions of the model.

To rephrase the introduction, the problem of binary n-string recognition can be solved by a dynamical system with trajectories that asymptotically approach the closest of the constant attractor trajectories (up to 2^n) in the orthants relative to T_ε. It is also possible to extend the memory model to include combinations of constant trajectories and limit cycles as memories (attractors).

From Chapter 1 let us recall the neural network model

$$dx_1/dt = -x_1 + g_1(1-g_2) - (1-g_1)g_2 \tag{1.2}$$
$$dx_2/dt = -x_2 - g_1(1-g_2) + (1-g_1)g_2$$

where each gain function is piecewise linear with three pieces (that is, a ramp function) with $\varepsilon = .1$. As mentioned in §2.6, (1.2) could be rewritten as

$$dx_1/dt = -x_1 + g_1 - g_2$$
$$dx_2/dt = -x_2 - g_1 + g_2$$

However, it would be a big mistake to try to analyze (1.2) and its generalizations in any such expanded algebraic form because only in the product form can one easily see which terms are nonzero in a particular orthant relative to T_ε. We illustrate typical trajectories for (1.2) by recalling Fig. 1.1 .

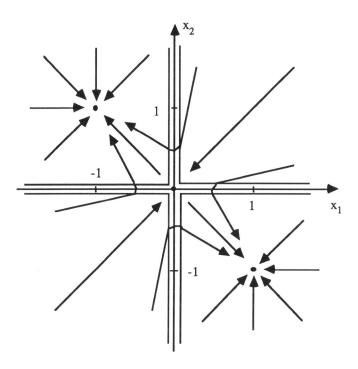

Figure 1.1. Typical trajectories for a memory model in two-dimensional space with two memories. The transition zone T_ε for the model is the "+" shaped region.

The system (1.2) has three constant trajectories: (-1,1), (0,0), and (1,-1). Except for the two special trajectories which asymptotically approach the unstable constant trajectory (0,0), nonconstant trajectories asymptotically approach either the constant trajectory (-1, 1) (the g vector for which is (0,1)) or the constant trajectory (1, -1) (the g vector for which is (1,0)). Thus (1.2) is an almost trivial system for generally deciding which of two memorized states is closer to a given state (used as the initial state of a trajectory).

A related system is

$$dx_1/dt = -x_1 + g_1(1-g_2) - (1-g_1)g_2 - (1-g_1)(1-g_2) \tag{3.3}$$
$$dx_2/dt = -x_2 - g_1(1-g_2) + (1-g_1)g_2 - (1-g_1)(1-g_2)$$

It is not difficult to see that this system has five constant trajectories: $(-1,1)$, $(-1,0)$, $(-1,-1)$, $(0,-1)$, and $(1,-1)$. The system (3.3) has certain trajectories which do not converge to the nearest memory, as shown in Fig. 3.7.

We also recall from Chapter 1

$$dx_1/dt = -x_1 - g_1g_2 - 2(1-g_1)g_2 + (1-g_1)(1-g_2) + 2g_1(1-g_2) \tag{1.3}$$
$$dx_2/dt = -x_2 + 2g_1g_2 - (1-g_1)g_2 - 2(1-g_1)(1-g_2) + g_1(1-g_2)$$

with dynamics as in Fig. 1.2.

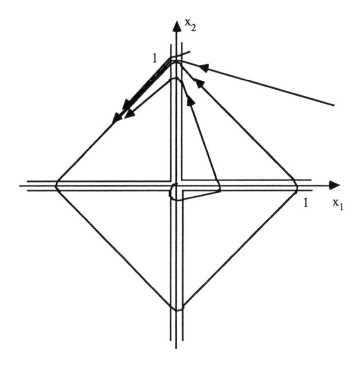

Figure 1.2. A neural network model with a limit cycle.

This model has exactly one attractor, a limit cycle. As explained in the following chapter, it is possible in spaces of higher dimension to specify an analogous model with several nonintersecting limit cycles as attractors. Any orthants relative to T_ε that do not contain any segment of a limit cycle can also contain constant attractors.

The models (1.2) and (1.3) are prototypes for the general memory model.

3.3 The memory model

We proceed to focus on a version of (3.1) which offers dense constant memory storage (cyclic memories will be developed in the next chapter). *Memories* are for our purposes binomial strings $\{\mathbf{m}_\iota\}$, $\iota = 1,...,$ M, having components $\mathbf{m}_{\iota j} = 1$ or 0, $j = 1,...,n$. We say $\mathbf{m}_\iota$ is *stored* as a constant trajectory $\underline{x}$ with components ± 1 if $g_j(\underline{x}_j) = \mathbf{m}_{\iota j}$. Neural networks used in content addressable memory or pattern recognition ideally should have only memories as attractors.

Suppose $\{\mathbf{m}_\iota\}$ are given memories. Let us form *image products* $\{\ I_\iota\ \}$ as

$$I_\iota = \prod_{j=1}^{n} \{\mathbf{m}_{\iota j} g_j(x_j) + (1-\mathbf{m}_{\iota j})(1-g_j(x_j))\} \tag{3.4}$$

For example, if n = 4 and $\mathbf{m}_1 = (0,1,0,1)$, then $I_1 = (1-g_1)g_2(1-g_3)g_4$ with associated constant trajectory $(-1, 1,-1, 1)$. Each I_ι satisfies $0 \le I_\iota \le 1$. As mentioned in Chapter 2, in the orthants relative to T_ε each I_ι is *turned on* $(= 1)$ in exactly one orthant and *turned off* $(= 0)$ in all others.

The version of (3.1) which we shall study is built with image products in (3.4) and is called the *memory model*

$$dx_i/dt = -x_i + \sum_{\iota=1}^{M} (2\mathbf{m}_{\iota i} -1)I_\iota(g) \tag{3.5}$$

It is quite possible to store many memories, even all 2^n, with much simpler models, namely versions of (3.1) with each p_i a linear sum of g components. (But generally such models have many spurious attractors.) As one might expect, the high order models are much more versatile. It can be shown, for example, that no three-dimensional model (3.1) using linear p_i functions has constant trajectories only in the orthants signed $(-,-,-)$, $(+,+,-)$,

(+,-,+), and (-,+,+) and trajectories invariant with respect to interchange of the three axes. As we proceed to show, the corresponding memory model (3.5) would certainly have these features.

We let **1** denote the vector (1, 1,..., 1) and **0** the vector (0, 0,..., 0).

THEOREM 3.1. If the state $\underline{x}$ is a constant trajectory for (3.5) in an orthant relative to T_ε, then $g(\underline{x})$ is a memory in the set $\{\mathbf{m}_\iota\}$. Furthermore, if $\underline{x}$ is defined in terms of a memory $\mathbf{m}_\iota$ by $\underline{x} = 2\mathbf{m}_\iota - \mathbf{1}$, then $\underline{x}$ is a stable constant trajectory for (3.5) with $g(\underline{x}) = \mathbf{m}_\iota$.

PROOF. Suppose $\underline{x}$ is a constant trajectory for (3.5) in an orthant relative to T_ε. Thus each component of $g(\underline{x})$ is 0 or 1. Each image product $I_\iota(g)$ is either 1 (if $g(\underline{x})$ is exactly memory $\mathbf{m}_\iota$) or 0, and at most one $I_\iota(g)$ can be 1. If every $I_\iota(g) = 0$, then $\underline{x}_i = \mathbf{0}$ and every $g_i = .5$, a contradiction. Thus exactly one $I_\iota(g)$ is on and $g(\underline{x})$ must be a memory.

Next choose one of the memories of (3.5), say, $\mathbf{m}_\iota$, and define state $\underline{x}$ by $\underline{x}_i = 2\mathbf{m}_{\iota i} - 1 = \pm 1$. Thus each $g_i(\underline{x}) = 0$ or 1 and $g(\underline{x})$ is exactly $\mathbf{m}_\iota$. Also, exactly one summand in Σ on the right side of (3.5) is nonzero, the ι summand $I_\iota(g(\underline{x})) = 1$. Plugging $\underline{x}$ into (3.5) yields $dx_i/dt = -\underline{x}_i + 2\mathbf{m}_{\iota i} - 1 = 0$, as required.   QED

We recall that if an orthant O relative to T_ε does not contain one of the memories in (3.5), then every trajectory while in O asymptotically approaches the origin of n-space. This approach continues until T_ε is entered. Hence selecting $\varepsilon = 0$ would lead for the differential equation version to an infinite time interval before escaping O.

A schematic representation of components of the memory model in (3.5) can be constructed as in Fig. 3.1. The state neuron itself is regarded as a capacitor and its state is characterized by a voltage. The neuron receives inputs of either sign and transmits outputs for processing. The $-x_i$ term in (3.5) can be thought of as charge leakage through a resistor. The symbols shown denote sign reversal interchange between g and 1-g. Diodes in the circuit prevent shorts. Realized as a schematic circuit, the system (3.5) has 5n edges associated with inputs and outputs of state neurons and M(1+2n) edges associated with the M memories.

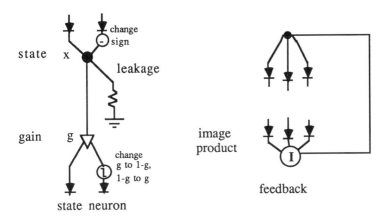

Figure 3.1. Schematic representation of components of the memory model (3.5).

For example, the memory model (3.5) with n = 4 and only one memory $m_1 = (0,1,0,1)$ is

$$dx_1/dt = -x_1 - (1-g_1)g_2(1-g_3)g_4 \qquad (3.6)$$
$$dx_2/dt = -x_2 + (1-g_1)g_2(1-g_3)g_4$$
$$dx_3/dt = -x_3 - (1-g_1)g_2(1-g_3)g_4$$
$$dx_4/dt = -x_4 + (1-g_1)g_2(1-g_3)g_4$$

By inspection, the only constant trajectory for (3.6) with g component values 0 or 1 is $\underline{x} = (-1,1,-1,1)$. All additional constant trajectories must lie in the transistion zone T_ε. The schematic circuit for (3.6) is shown in Fig. 3.2.

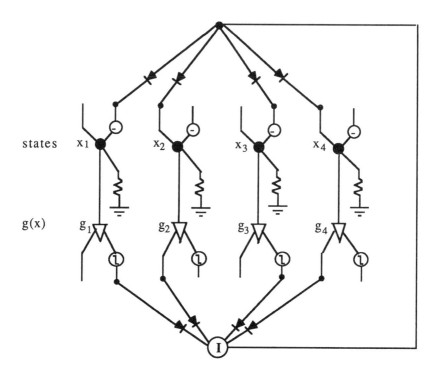

Figure 3.2. The schematic circuit corresponding to system (3.6), a 4-neuron system with only one memory. Initial input to x and final output from g are not shown. As suggested by this schematic, the memory model neural network can be regarded as a fully parallel analog computer with as many feedback loops as memories.

3.4 Dynamics of the Memory Model

In this section we assume that all ε_i are equal, $\varepsilon_i = \varepsilon$, and for the moment that each gain function is a ramp function. Each point x in T_ε can be characterized by the number of coefficients of x with $|x_i| < \varepsilon$, that is, the number of neurons *in transition*. If exactly one neuron is in transistion, say, $|x_1| < \varepsilon$ and $x = (x_1, x)$ (using x to denote the other coefficients of the vector x), then there are four possible trajectory types in the two *adjacent orthants* (the orthants containing $(1, x)$ and $(-1, x)$. If neither adjacent orthant contains a memory (constant attractor with coeffcients ±1), then the trajectories near x are as depicted in Fig. 3.3.

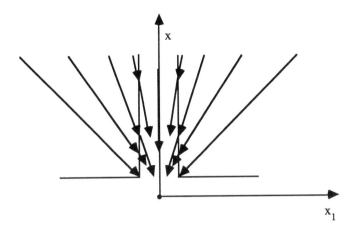

Figure 3.3. Trajectories in a portion of T_ε between two orthants without memories. This figure (and subsequent figures) is a two-dimensional projection. Actual trajectories proceed along n-dimensional rays toward the origin until some portion of T_ε is reached with more neurons in transition.

If exactly one orthant contains a memory, then the trajectories near x are as depicted in Fig. 3.4.

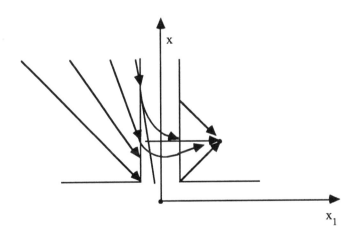

Figure 3.4. Trajectories in a portion of T_ε between two orthants, exactly one with a memory. Note the relative position of trajectory rays extended into T_ε and typical trajectories.

If both adjacent orthants contain memories, then the trajectories near x are as depicted in Fig. 3.5.

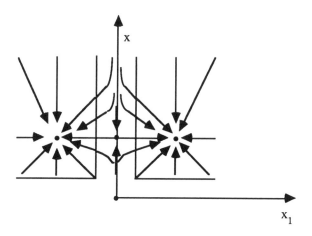

Figure 3.5. Trajectories in a portion of T_ε between two orthants, each with a memory.

Now consider the case that x lies in a portion of T_ε with exactly two neurons in transition. Near x and in the four adjacent orthants, several trajectory patterns are possible. A pattern in which three of the orthants contain attractors is shown in Fig. 3.6.

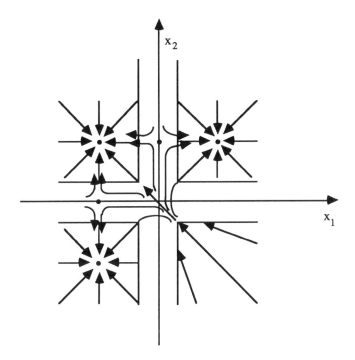

Figure 3.6. Trajectories in a portion of T_ϵ with two neurons in transition. Four adjacent orthants are shown, three of which have memories.

These figures suggest some general features of constant trajectories in T_ϵ which we next develop. We refer to the 2^n points in n-space with coefficients ± 1 as the *vertices of the n-cube* and are the potential representations of memories of the memory model. Naturally, if we hold constant n-p of these coefficients, the remaining 2^p vertices constitute vertices of a p-cube which is embedded in the n-cube for $1 \le p \le n$. A vertex in such a p-cube thus has n-p coefficients in common with all other vertices in the p-cube, its *fixed* coefficients. In particular, fixing all but the first coefficient yields a 1-cube with vertices $(+1, x)$ and $(-1, x)$ (again using x to denote the other coefficients of the n-vector x). We call a set of q $(2 \le q \le 2^p)$ vertices in such a p-cube *symmetric* if the sum of the vertices as vectors is $(0, qx)$. Suppose $(+1, x)$ and $(-1, x)$ represent memories of a memory model, that is, they are associated with the memories (binary strings) $\mathbf{m_1} = (1, m)$ and $\mathbf{m_2} = (0, m)$. Let us write the associated image products $\mathbf{I_1}$ and $\mathbf{I_2}$. Let us denote the product of the common terms in the two image products by $\mathbf{J}(g)$, so $\mathbf{I_1} = g_1\mathbf{J}(g)$ and $\mathbf{I_2} = (1-g_1)\mathbf{J}(g)$. Since the two vertices are symmetric vertices in a

1-cube, we have for each j > 1, $m_{1j} = m_{2j}$ with $x_j = 2\, m_{1j} -1 = 2\, m_{2j} -1$. At $(0, x)$ we have $J(g) = 1$ and so

$$dx_1/dt = -0 + .5\, J(g) - (1-.5)J(g) = 0$$
$$dx_j/dt = -x_j + .5(2\, m_{1j} -1)\, J(g) + .5(2\, m_{2j} -1)\, J(g) = 0$$

It follows that $(0, x)$ is also a constant trajectory of the type depicted in T_ε in Fig. 3.6. This observation generalizes as follows.

THEOREM 3.2. Suppose q ≤ 2p of the memories of a memory model are symmetric vertices of a p-cube, written for convenience as $(\underline{x}, x)$ where x is the fixed part of the memories. Then $(0, q(.5)^p x)$ is also a constant trajectory for the model.

PROOF. This time we form $J(g)$ by deleting the first p of the terms in the q associated image products $I_1 ,..., I_q$. For each j > p, $m_{1j} = ... = m_{qj}$ with $x_j = 2\, m_{1j} -1 = ... = 2\, m_{qj} -1 = \pm 1$. At $(0, q(.5)^p x)$ we have $J(g) = 1$ and so for i = 1, ..., p and j > p

$$dx_i/dt = -0 + q(.5)^p\, J(g) - q(.5)^p\, J(g) = 0$$
$$dx_j/dt = - q(.5)^p x_j + (.5)^p(2\, m_{1j} -1)\, J(g) +... +(.5)^p(2\, m_{qj} -1)\, J(g) = 0 \qquad \text{QED}$$

We note that $\varepsilon < .5^n$ implies that for a constant trajectory $(0, q(.5)^p x) = (0, ..., 0, \pm q(.5)^p, ..., \pm q(.5)^p)$ as obtained in Theorem 4, the g vector is some $(.5, ..., .5, \pm 1, ..., \pm 1)$. In particular, if p = n and all memories correspond to symmetric vertices of the n-cube, then the origin 0 is a constant trajectory with $g = (.5, ..., .5)$.

Some versions of the memory model have unstable constant trajectories in T_ε which are not of the type in Theorem 3.6. For example, the three-dimensional model which stores memories (1,1,1), (1,0,0), (0,1,0), and (0,0,1) as symmetric vertices (1,1,1), (1,-1,-1), (-1,1,-1), and (-1,-1,1) has constant trajectories in T_ε as shown in Fig. 3.7.

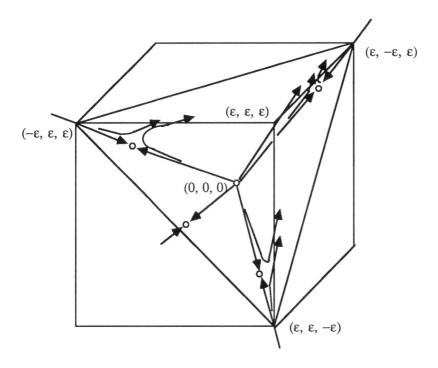

Figure 3.7. Representative trajectories for the memory model with constant trajectories (1,1,1), (1,-1,-1), (-1,1,-1), and (-1,-1,1) as memories. Shown is the portion of T_ε in which all three neurons are in transition. Note the existence of unstable constant trajectories (open circles). If $\varepsilon - 2\varepsilon^2$ is denoted by α, then the unstable constant trajectories occur at $(-\alpha, \alpha, \alpha)$, $(\alpha, -\alpha, \alpha)$, $(\alpha, \alpha, -\alpha)$, and at the origin.

We now modify our model slightly by choosing a new type of gain function. Let δ be a positive constant, $\delta < .5$, and for $|x_i| < \varepsilon$ define $g_i(x_i) = .5 + x_i(1-2\delta)/(2\varepsilon)$ (see Fig. 3.8). This gain function is unconventional but lends itself to particularly straightforward analysis. Simulations indicate using instead a conventional ramp (piecewise linear) gain function or smooth sigmoidal gain function in (3.5) does not lead to significantly different dynamics.

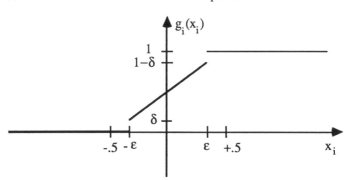

Figure 3.8. A type of gain function which simplifies stability analysis.

THEOREM 3.3. If in addition to the above specifications we require ε to satisfy $g_i'(0) = (1-2\delta)/(2\varepsilon) > n\delta^{1-n}$, then any constant trajectory for (5) in T_ε is unstable.

PROOF. Let the product $J_{\iota i}$ formed by deleting the g_i term from I_ι, that is, $J_{\iota i}$ $= I_\iota/\{(m_{\iota i}g_i+(1- m_{\iota i})(1-g_i)\}$. The ii entry in the linear approximation matrix L of (3.5) about any constant trajectory $\underline{x}$ with $\underline{x}_i \neq \pm\varepsilon$ for each $i = 1,2,..., n$ is L_{ii} $= -1 + \Sigma_\iota J_{\iota i} g_i'$ where $\iota = 1...M$. Each g_i' in this sum must be replaced with 0 or $(1-2\delta)/(2\varepsilon)$, depending upon the region of T_ε in which the linear approximation is to hold. Note that Σ summands in L_{ii} are all nonnegative.

Suppose all $I_\iota = 0$ at an infeasible constant trajectory $\underline{x}$ in T_ε. Then all $\underline{x}_i$ $= 0$, so all $I_\iota = (.5)^n$, a contradiction. Suppose $I_\iota = 1$. Then all other $I_k = 0$, all $\underline{x}_i = \pm 1$, and $g(\underline{x})$ is actually a memory, a contradiction. Thus some $\{g_j(\underline{x}_j)\}$ must satisfy $\delta < g_j(\underline{x}_j) < 1- \delta$ and some ι exists with $J_{\iota j} \neq 0$, that is, $J_{\iota j} > \delta^{n-1}$. Hence some $L_{jj} = -1 + \Sigma_\iota J_{\iota j} g_j' > -1 + \delta^{n-1} (1-2\delta)/(2\varepsilon) > -1 + n$. Hence the trace of L is positive. This is a brute force way of guaranteeing that at least one eigenvalue of L has positive real part and so $\underline{x}$ is unstable.   QED

Figures 3.3-3.7 suggest that a trajectory for a memory model starting in any orthant relative to T_ε generally converges to the nearest memory. As discussed in the next chapter, this characteristic has been statistically established by a large number of computer simulations [JP]. However, as the astute reader might have noticed, there are trajectories in Fig. 3.6 which do not

converge to the closest memory. In Fig. 3.6 the trajectory starting at (1,-1)
converges to (-1,1), as do other trajectories starting near (1,-1). This situation
is improved by using a smaller ε, as suggested in Fig. 3.9 for the related model
given explicitly by equations (3.3)

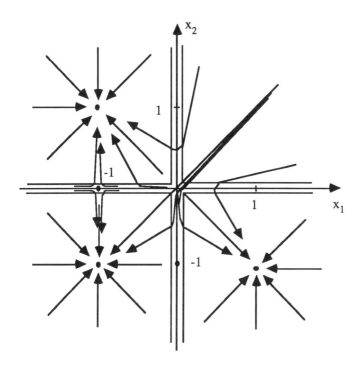

Figure 3.9. Typical trajectories for the memory model (3.3) in two-dimensional
space with three memories. The transition zone $T_ε$ for the model is the "+"
shaped region. Note the trajectories starting at (1,1) and near (1,1).

As shown in Fig. 3.9, the trajectory starting exactly at (1,1) passes
through (0,0) and asymptotically approaches (-1,-1), even though (1,1) is
closer to the two other attractors. Furthermore, since trajectories depend
continuously upon initial conditions, there is a corridor of trajectories starting
near (1,1) which make the same error. However, it can be shown that
choosing ε suitably small makes this corridor arbitrarily narrow.

We also note that a trajectory of the general memory model (3.5) which
starts in an orthant containing no $m_ι$ satisfies $dx_i/dt = -x_i$, that is, consists of
a straight line trajectory which asymptotically approaches the origin until

entering T_ε after a finite time interval. Roughly speaking, a small ε as required by Theorem 3.3 would imply tight packing and sharp turns for such trajectories near the origin. These aspects, while no trouble for the mathematical system (3.5), might amount to precision difficulties for actual microelectronic circuits. On the other hand, one might make use of the fact that a trajectory with an initial state very nearly equidistant (in terms of convergence) to two memories would linger near the origin. That is, one might detect and discard an ambiguous initial state by setting a time limit on convergence to a memory.

3.5 Modified Memory Models Using Foundation Functions

We recall from Chapter 2 that certain high order memory models can be developed in terms of foundation functions.

THEOREM 2.3. Suppose we are given a neural network model of the form

$$dx_i/dt = -x_i - \lambda_i^{-1}\partial\Phi/\partial g_i$$

for n positive constants $\{\lambda_i\}$ and a continuously differentiable function Φ. Then every trajectory for the model must be or must asymptotically approach a constant trajectory.

Suppose $\{m_\iota\}$ are given memories. Using the image products $\{I_\iota\}$ in (3.4), that is,

$$I_\iota = \prod_{j=1}^{n}\{m_{\iota j}\,g_j(x_j)+(1-m_{\iota j})(1-g_j(x_j))\} \tag{3.4}$$

let us define

$$\Phi = .5 \left\{ \prod_{j=1}^{n}\{m_{\iota j}\,g_j(x_j)+(1-m_{\iota j})(1-g_j(x_j))\} \right\}^2 \tag{3.7}$$

Using all $\lambda_i = 1$ in Theorem 2.3 leads to the following *modified memory model*:

$$dx_i/dt = -x_i + \sum_{\iota=1}^{M} (2m_{\iota i} - 1) I_\iota(g) J_{\iota i}(g) \qquad (3.8)$$

where the product $J_{\iota i}$ is formed by deleting the g_i term from I_ι, that is, $J_{\iota i} = I_\iota / \{(m_{\iota i} g_i + (1 - m_{\iota i})(1 - g_i)) \}$.

For example, if $n = 4$ and we wish to store only one memory $m_1 = (0,1,0,1)$, then we use $I_1 = (1-g_1)g_2(1-g_3)g_4$ and $\Phi = .5\{(1-g_1)g_2(1-g_3)g_4\}^2$. Also

$$dx_1/dt = -x_1 - (1-g_1)g_2^2(1-g_3)^2 g_4{}^2 \qquad (3.9)$$
$$dx_2/dt = -x_2 + (1-g_1)^2 g_2(1-g_3)^2 g_4{}^2$$
$$dx_3/dt = -x_3 - (1-g_1)^2 g_2^2(1-g_3)g_4{}^2$$
$$dx_4/dt = -x_4 + (1-g_1)^2 g_2^2(1-g_3)^2 g_4$$

Unfortunately, it seems difficult to represent even this simple system graphically. But this modified memory model does fit into the foundation function format.

We proceed to characterize the constant trajectories of (3.8) in the orthants relative to T_ε.

THEOREM 3.4. If the state $\underline{x}$ is a constant trajectory for (3.8) in an orthant relative to T_ε, then $g(\underline{x})$ is a memory in $\{m_\iota\}$. Furthermore, any $\underline{x}$ defined in terms of memory m_ι by $\underline{x}_i = 2m_{\iota i} - 1$ is a stable constant trajectory for (3.8).

PROOF. Suppose $\underline{x}$ is a constant trajectory for (3.8) in an orthant relative to T_ε. Thus each component of $g(\underline{x})$ is 0 or 1. Each image product $I_\iota(g)$ and each product $I_\iota(g)J_{\iota i}(g)$ is either 1 (if $g(\underline{x})$ is exactly memory m_ι) or 0, and at most one $I_\iota(g)J_{\iota i}(g)$ can be 1. Also, $2m_{\iota i} - 1$ is either $+1$ (if $m_{\iota i}$ is 1) or -1 (if $m_{\iota i}$ is 0). Thus for each sum Σ in (3.7) to be nonzero (so each $|\underline{x}_i| \geq \varepsilon$), $g(\underline{x})$ must be a memory.

Next choose one of the memories of (3.8), say, m_1, and define state $\underline{x}$ by $\underline{x}_i = 2m_{1i} - 1$. Since $\varepsilon < 1$, each $g_i(\underline{x}) = 0$ or 1 and $g(\underline{x})$ is exactly m_1. Thus exactly one summand in Σ on the right side of (3.8) is nonzero, the $\iota = 1$ summand with $I_1(g)J_{1i}(g) = 1$. Plugging $\underline{x}$ into (3.8) yields

$$dx_i/dt \text{ at } \underline{x} = -\underline{x}_i + 2m_{1i} - 1 = 0 \qquad \text{QED}$$

A second class of memory models built using foundation functions can be described as follows. Let

$$\Phi = \prod_{j=1}^{n} \{ m_{\iota j} g_j(x_j) + (1 - m_{\iota j})(1 - g_j(x_j)) \} \tag{3.10}$$

Using all $\lambda_i = 1$ in Theorem 2.3 leads to the following modification of the memory model:

$$dx_i/dt = -x_i + \sum_{\iota=1}^{M} (2m_{\iota i} - 1) J_{\iota i}(g) \tag{3.11}$$

where again the product $J_{\iota i}$ is formed by deleting the g_i term from I_ι , that is, $J_{\iota i} = I_\iota / \{ (m_{\iota i} g_i + (1 - m_{\iota i})(1 - g_i) \}$.

THEOREM 3.5. Suppose any distinct memories differ in at least two components. If the state $\underline{x}$ is a constant trajectory for (3.11) in an orthant relative to T_ε , then $g(\underline{x})$ is a memory in $\{ m_\iota \}$. Furthermore, any $\underline{x}$ defined in terms of memory m_ι by $\underline{x}_i = 2m_{\iota i} - 1$ is a stable constant trajectory for (3.11).

PROOF. Suppose $\underline{x}$ is a constant trajectory for (3.11) in an orthant relative to T_ε. Thus each component of $g(\underline{x})$ is 0 or 1. Consider one of the products $J_{\iota i}(g)$. If each $g_j(\underline{x})$ except possibly $g_i(\underline{x})$ is exactly $m_{\iota j}$, then the product $J_{\iota i}(g)$ is 1; otherwise the product $J_{\iota i}(g)$ is 0. Since two distinct memories differ in at least two components, at most one $J_{\iota i}(g)$ is 1. Since $\underline{x}_i \neq 0$, at least one such product $J_{\iota i}(g) \neq 0$. Thus exactly one $J_{\iota i}(g)$ at $\underline{x}$ is 1. Also, $2m_{\iota i} - 1$ is either +1 (if $m_{\iota i}$ is 1) or -1 (if $m_{\iota i}$ is 0). Thus each $\underline{x}_i = \pm 1$ and $g(\underline{x})$ must be the memory m_ι .
   Next choose one of the memories of (3.11), say, m_1, and define state $\underline{x}$ by $\underline{x}_i = 2m_{1i} - 1$. Since $\varepsilon < 1$, each $g_i(\underline{x}) = 0$ or 1 and so $g(\underline{x})$ is exactly m_1 .
Considereing the Hamming distance between any two memories, exactly one summand in Σ on the right side of (3.11) is nonzero, the $\iota = 1$ summand with $J_{1i}(g) = 1$. Plugging $\underline{x}$ into (3.11) yields

$$dx_i/dt \text{ at } \underline{x} = -\underline{x}_i + 2m_{1i} - 1 = 0 \hspace{2cm} \text{QED}$$

For the sake of comparison, the (3.11) analog of the model (3.6) with $n = 4$ and only one memory $\mathbf{m_1} = (0,1,0,1)$ uses

$$\Phi = (1-g_1)g_2(1-g_3)g_4$$

to obtain

$$dx_1/dt = -x_1 - g_2(1-g_3)g_4 \qquad\qquad (3.12)$$
$$dx_2/dt = -x_2 + (1-g_1)(1-g_3)g_4$$
$$dx_3/dt = -x_3 - (1-g_1)g_2g_4$$
$$dx_4/dt = -x_4 + (1-g_1)g_2(1-g_3)$$

Again, this alternative to the memory model (3.6) seems difficult to represent graphically, in contrast to Fig. 3.2. However, (3.12) is built from a foundation function and uses lower order multinomials than (3.6).

Suppose in three-dimensional space we try to store memories $(1,1,1)$, $(1,0,0)$, $(0,1,0)$, and $(0,0,1)$. The model (3.11) takes the form

$$dx_1/dt = -x_1 + g_2g_3 +(1-g_2)(1-g_3)-g_2(1-g_3)-(1-g_2)g_3$$
$$dx_2/dt = -x_2 + g_1g_3 -g_1(1-g_3)+(1-g_1)(1-g_3)-(1-g_1)g_3$$
$$dx_3/dt = -x_3 + g_1g_2-g_1(1-g_2)-(1-g_1)g_2+(1-g_1)(1-g_2)$$

At the origin, the linear approximation matrix of this system is -1 times the identity matrix, having all eigenvalues -1 regardless of ε. Thus for this model no choice of ε can destabilize the infeasible constant trajectory at the origin.

3.6 Comparison of the Memory Model and the Hopfield Model

The *Hopfield model* [Ho], as used by Hopfield and others, is a neural network model of the form

$$dx_i/dt = -x_i - \lambda_i^{-1}\partial\Phi/\partial g_i$$
$$dx_i/dt = I_i - x_i + \sum_{j=1}^{n} T_{ij} g_j(x_j) \qquad\qquad (3.13)$$

where T_{ij} is a symmetric matrix.   In terms of a foundation function we could use

$$\Phi = -\sum_{i=1}^{n} I_i g_i \ -\ .5 \sum_{i,j=1}^{n} T_{ij} g_i g_j$$

and apply Theorem 2.3. Note that we need not require T_{ij} to be symmetric if constants (Volterra multipliers) $\{\lambda_i\}$ exist with $\lambda_i T_{ij} = \lambda_j T_{ji}$.

There are some similarities between the memory model (3.5) and the Hopfield model (3.13). It is interesting to note that if we wish to store all 2^n possible memories in n-space, then the memory model algebraically reduces to a Hopfield model, namely

$$dx_i/dt = -1 - x_i + 2g_i$$

At the other exteme of memory density, we can compare the memory model and the linear model which store in two-dimensional space the two memories $(0,1)$ and $(1,0)$ at the points $(-1,1)$ and $(1,-1)$. The dynamics of the resulting memory model and Hopfield model can be compared as follows. The memory model is

$$dx_1/dt = -x_1 + g_1(1-g_2) - (1-g_1)g_2 \tag{3.2}$$
$$dx_2/dt = -x_2 - g_1(1-g_2) + (1-g_1)g_2$$

while the Hopfield model with the same attractors is

$$dx_1/dt = 1 - x_1 - 2g_2$$
$$dx_2/dt = 1 - x_2 - 2g_1$$

The dynamics of these models are shown in Fig. 3.10.

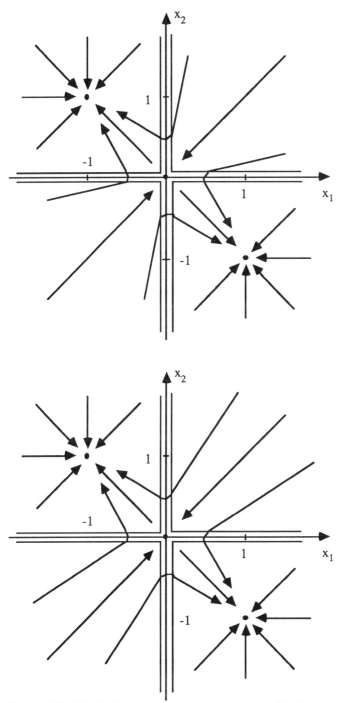

Figure 3.10. Typical trajectories for a memory model (above) and a Hopfield model in two-dimensional space with the two memories (0,1) and (1,0).

As can be seen in Fig. 3.10, the memory model and the linear model are qualitatively very similar.

However, suppose we try to devise a linear model (3.13) with $n = 2$ which stores exactly three of the four binary two-strings, namely (1,1), (1,0), and (0,1) at the points (1,1), (1,-1), and (-1,1). Suppose we also require that the trajectories of the system be symmetric with respect to the line $x_1 = x_2$, as are the trajectories in Fig. 3.10 . Thus we requrire $I_1 = I_2 = I$, $T_{11} = T_{22} = D$ and $T_{12} = T_{21} = T$. To avoid a constant trajectory in the (-,-) orthant, such a model would need I positive. To have (1,-1) as a constant trajectory, $D + I = 1$ and $T + I = -1$. To have (1,1) as a constant trajectory, $D + I + T = 1$, so $T = 0$; but then $I = -1$, a contradiction. Thus the linear model and the memory model are not equivalent in terms of freedom of specification of constant trajectories.

One might ask nonetheless for a linear model with given memories (g values) but corresponding x coordinates not necessarily $= \pm 1$. In three-dimensional space, we might try to store as memories

$x = (-1,-1,-1)$ with g = (0,0,0)
$x = (1,1,-1)$ with g = (1,1,0)
$x = (1,-1,1)$ with g = (1,0,1)
$x = (-1,1,1)$ with g = (0,1,1)

as constant trajectories with a linear system with coordinate symmetry like that of the memory model with the same memories. Thus we require in (3.13) that $I_i = I$, $T_{ii} = D$, and $T_{ij} = T_{ji} = T$ for i,j = 1,2,3 and $i \neq j$. Solving for positive and negative x values for such memories leads to the following conditions:

(1) store (0,0,0) $\Rightarrow I < 0$
(2) not store (1,0,0) , (0,1,0), and (0,0,1) $\Rightarrow D < -I$ or $-I < T$
(3) store (1,1,0) , (1,0,1), and (0,1,1) $\Rightarrow 2T < -I < D+T$
(4) not store (1,1,1) $\Rightarrow D+2T < -I$

Here (4) and (3) imply $D+2T < D+T$, that is, T is negative. Since T is negative and, by (1), -I is positive, (2) implies $D < -I$. Then $D+T < -I+T < -I$, contradicting (3). Thus no linear system stores only the given memories and is otherwise symmetric with respect to x_1, x_2, and x_3 .

PROBLEMS

1. Devise the memory model with stores as memories

$x = (-1,-1,-1)$ with $g = (0,0,0)$
$x = (1,1,-1)$ with $g = (1,1,0)$
$x = (1,-1,1)$ with $g = (1,0,1)$
$x = (-1,1,1)$ with $g = (0,1,1)$

2. Analyze

$dx_1/dt = -x_1 + g_2g_3 + (1-g_2)(1-g_3) - g_2(1-g_3) - (1-g_2)g_3$
$dx_2/dt = -x_2 + g_1g_3 - g_1(1-g_3) + (1-g_1)(1-g_3) - (1-g_1)g_3$
$dx_3/dt = -x_3 + g_1g_2 - g_1(1-g_2) - (1-g_1)g_2 + (1-g_1)(1-g_2)$

3. Analyze

$dx_1/dt = -x_1 - (1-g_1)g_2^2(1-g_3)^2g_4^2 - (1-g_1)g_2^2(1-g_3)^2(1-g_4)^2$
$dx_2/dt = -x_2 + (1-g_1)^2g_2(1-g_3)^2g_4^2 + (1-g_1)^2g_2(1-g_3)^2(1-g_4)^2$
$dx_3/dt = -x_3 - (1-g_1)^2g_2^2(1-g_3)g_4^2 - (1-g_1)^2g_2^2(1-g_3)(1-g_4)^2$
$dx_4/dt = -x_4 + (1-g_1)^2g_2^2(1-g_3)^2g_4 - (1-g_1)^2g_2^2(1-g_3)^2(1-g_4)$

ANSWERS

1. The appropriate memory model is

$$dx_1/dt = -x_1 - (1-g_1)(1-g_2)(1-g_3) + g_1g_2(1-g_3) + g_1(1-g_2)g_3 - (1-g_1)g_2g_3$$
$$dx_2/dt = -x_2 - (1-g_1)(1-g_2)(1-g_3) + g_1g_2(1-g_3) - g_1(1-g_2)g_3 + (1-g_1)g_2g_3$$
$$dx_3/dt = -x_3 - (1-g_1)(1-g_2)(1-g_3) - g_1g_2(1-g_3) + g_1(1-g_2)g_3 + (1-g_1)g_2g_3$$

2. This modified memory model stores

$x = (1,1,1)$ with $g = (1,1,1)$
$x = (1,-1,-1)$ with $g = (1,0,0)$
$x = (-1,1,-1)$ with $g = (0,1,0)$
$x = (-1,-1,1)$ with $g = (0,0,1)$

as constant attractors. The origin is an unstable constant trajectory. If $\varepsilon = .1$, then $(.01,.01,.01)$, $(.01,-.01,-.01)$, $(-.01,.01,-.01)$, $(-.01,-.01,.01)$ are also unstable constant trajectories. The performance of this system is qualitatively much like that of the memory model with the same memories. Note that the memories differ pairwise in two components. A spreadsheet version of this model is shown on page 80.

3. This modified memory model stores

$x = (-1,1,-1,1)$ with $g = (0,1,0,1)$
$x = (-1,1,-1,-1)$ with $g = (0,1,0,0)$

as constant attractors. Another constant trajectory is

$y = (-.5,.5,-.5,0)$ with $g = (0,1,0,.5)$

Assuming all $\varepsilon_i = \varepsilon$, $g' = (2\varepsilon)^{-1}$, the diagonal entries in linear approximation matrix L at y are $-1+.5(2\varepsilon)^{-1}$, $-1+.5(2\varepsilon)^{-1}$, $-1+.5(2\varepsilon)^{-1}$, and -1. Thus the trace of L is $-4+1.5(2\varepsilon)^{-1}$. If $\varepsilon < .1875$, the trace of L is positive, implying that at least one eigenvalue of L is positive and y is unstable. It can be shown by other means that y is also unstable for $\varepsilon \geq .1875$. The performance of this system is qualitatively much like that of the memory model with the same memories. Note that the memories differ pairwise in two components. A spreadsheet version of this model is shown on page 81.

	A	B	C	D	E
1	delta t =	0.1			epsilon =
2	rand start =	0.8528175			0.1
3	start cell =	0.9967198			
4					
5	x =	-0.996068			
6		-0.995303			
7		0.9967198			
8					
9	g(x) =	0			
10		0			
11		1			
12					
13	next x =	-0.9964612			
14		-0.9957727			
15		0.9970478			
16					
17	loop =	-0.9964612			
18		-0.9957727			
19		0.9970478			
20					
21					
22					
23					
24					
25					
26					
27					
28					
29					
30					

	A	B	C	D	E
1	delta t =	0.1			epsilon =
2	rand start =	-0.4991699			0.1
3	start cell =	-0.9999505			
4					
5	x =	-0.9999734			
6		0.9999734			
7		-0.9999734			
8		-0.9999505			
9					
10	g(x) =	0			
11		1			
12		0			
13		0			
14					
15	next x =	-0.9999761			
16		0.9999761			
17		-0.9999761			
18		-0.9999554			
19					
20	loop =	-0.9999761			
21		0.9999761			
22		-0.9999761			
23		-0.9999554			
24					
25					
26					
27					
28					
29					
30					

Chapter 4--Code Recognition, Digital Communications, and General Recognition

4.1 Error Correction for Binary Codes

The memory model of Chapter Three can be used to solve the problem of code recognition. Suppose M binomial n-strings are chosen with components selected from $\{0,1\}$ (or equivalently from $\{-1,1\}$). This amounts to choosing M vertices $\{m_\iota\}$, $\iota = 1,2,..., M$, of the 2^n vertices of the (0,1)-binomial cube in n-dimensional space. Each chosen vertex is called a *code word*.

There are two objectives in designing an error-correcting code. The first, given n and M, is to choose the M code words as far apart as possible, using Hamming distance. (The *Hamming distance* between two binomial strings is the number of coefficients in which the strings differ.) The minimum distance between all pairs of words in a code is denoted by d. Making such a choice is called *encoding*. The second objective is to devise an algorithm for going from some vertex not in $\{m_\iota\}$ (representing a code word with error) to the nearest vertex in $\{m_\iota\}$ (the corrected code word). Using such an algorithm is called *decoding* [MS,P].

Some codes are labeled by the pair (n, $\log_2$ M), for example the *Hamming (7,4) code*. The Hamming (7,4) code is called a linear code because the conventional decoding algorithm entails linear binomial arithmetic. The Hamming (7,4) code with n = 7, M = 16 = 2^4, and d = 3 [P p. 3] appears in Table 4.1.

word to be encoded	binomial string							memory label
0	0	0	0	0	0	0	0	m_1
1	0	0	0	1	1	1	1	m_2
2	0	0	1	0	1	1	0	m_3
3	0	0	1	1	0	0	1	m_4
4	0	1	0	0	1	0	1	m_5
5	0	1	0	1	0	1	0	m_6
6	0	1	1	0	0	1	1	m_7
7	0	1	1	1	1	0	0	m_8
8	1	0	0	0	0	1	1	m_9
9	1	0	0	1	1	0	0	m_{10}
10	1	0	1	0	1	0	1	m_{11}
11	1	0	1	1	0	1	0	m_{12}
12	1	1	0	0	1	1	0	m_{13}
13	1	1	0	1	0	0	1	m_{14}
14	1	1	1	0	0	0	0	m_{15}
15	1	1	1	1	1	1	1	m_{16}

Table 4.1. The Hamming (7,4) code. Note all pairs of code words differ in at least three entries; some differ in four.

Since $d = 3$, a code word with one error is closer to the correct code word than any other code word. The linear decoding algorithm for the Hamming (7,4) code is given in [P p. 4, 137]. Alternatively, decoding can be accomplished using the memory model (3.5) with given memories $\{m_i\}$. We have tested simulations of the seven-dimensional memory model with such memories using $\varepsilon = .01$ and an Euler differencing scheme with $\Delta t = .5$. Convergence from a random point with coefficients of magnitude less than 1 to an orthant containing a code word generally occurs in fewer than 10 time steps. If we choose a code word at random, reverse exactly one of its seven digits, and feed that ± 1 binary string to the memory model as an initial state, then convergence to the original code word, that is, correction of one error, occurs in about five time steps.

The memory model can also be applied to the nonlinear Hadamard code with $n = 11$, $M = 24$, and $d = 5$ (for conventional decoding see [MS p. 39]). The Hadamard code is shown in Table 4.2. Simulations using $\varepsilon = .001$ result in correction of one error in about ten time steps and correction of two errors in about twelve time steps.

word to be encoded	binomial string											memory label
0	0	0	0	0	0	0	0	0	0	0	0	m_1
1	1	1	0	1	1	1	0	0	0	1	0	m_2
2	0	1	1	0	1	1	1	0	0	0	1	m_3
3	1	0	1	1	0	1	1	1	0	0	0	m_4
4	0	1	0	1	1	0	1	1	1	0	0	m_5
5	0	0	1	0	1	1	0	1	1	1	0	m_6
6	0	0	0	1	0	1	1	0	1	1	1	m_7
7	1	0	0	0	1	0	1	1	0	1	1	m_8
8	1	1	0	0	0	1	0	1	1	0	1	m_9
9	1	1	1	0	0	0	1	0	1	1	0	m_{10}
10	0	1	1	1	0	0	0	1	0	1	1	m_{11}
11	1	0	1	1	1	0	0	0	1	0	1	m_{12}
12	0	0	1	0	0	0	1	1	1	0	1	m_{13}
13	1	0	0	1	0	0	0	1	1	1	0	m_{14}
14	0	1	0	0	1	0	0	0	1	1	1	m_{15}
15	1	0	1	0	0	1	0	0	0	1	1	m_{16}
16	1	1	0	1	0	0	1	0	0	0	1	m_{17}
17	1	1	1	0	1	0	0	1	0	0	0	m_{18}
18	0	1	1	1	0	1	0	0	1	0	0	m_{19}
19	0	0	1	1	1	0	1	0	0	1	0	m_{20}
20	0	0	0	1	1	1	0	1	0	0	1	m_{21}
21	1	0	0	0	1	1	1	0	1	0	0	m_{22}
22	0	1	0	0	0	1	1	1	0	1	0	m_{23}
23	1	1	1	1	1	1	1	1	1	1	1	m_{24}

Table 4.2. The Hadamard code with n = 11, M = 24. Pairs of code words differ in at least five components.

4.2 Additional Tests of the Memory Model as a Decoder

A measure of noise on a communications channel is the *signal to noise ratio*. Suppose

n = the number of bits per code word (symbol)
0, v = the voltages of the 0 and 1 of the signal
α = the average number of 1's per code word

Suppose also that noise means a voltage added to each bit and that noise voltage is uniformly randomly distributed over the range [-a, +a] . Electric power flowing through a resistor (constant resistance) is proportional to the square of voltage. It can be shown that the average of the square of a long sequence of variables uniformly randomly distributed over the range [-a, a] is $(1/3)a^2$. Thus the average ratio of signal power to noise power is $3\alpha v^2/(na^2)$. A useful term derived from this power ratio is the *signal to noise ratio for noise uniformly distributed over* [-a, a], namely

$$10 \log_{10} \left(\frac{3 \alpha v^2}{n a^2} \right)$$

This number in decibels (dB) is a convenient expression for the quality of transmission.

Similarly, the signal to noise ratio for Gaussian noise can be defined. Suppose the probability that the added noise voltage is between x_1 and x_2 is

$$P(x) = \frac{1}{\sqrt{2\pi}\,\sigma} \int_{x_1}^{x_2} \exp\left[-\frac{x^2}{2\sigma^2} \right] dx$$

where σ is a positive constant in volts and σ^2 is the *variance* of the distribution. It can be shown that the mean square value of the noise is σ^2 [L 188]. This leads to a *signal to noise ratio for Gaussian noise with variance σ^2* of

$$10 \log_{10} \left(\frac{\alpha v^2}{n \sigma^2} \right)$$

P. Protzel has conducted tests [JP, P] using some standard codes with code words perturbed by Gaussian noise of various signal to noise ratios. He reports that memory model decoding using soft (nonbinary) input can result in fewer errors than conventional decoders which must first round input to binary form. Thus a dedicated electronic or optical integrated circuit based on something like the memory model design might be effective at high speed code correction.

The performance of the memory model as a decoder can be seen in the following graphs which summarize billions of simulations.

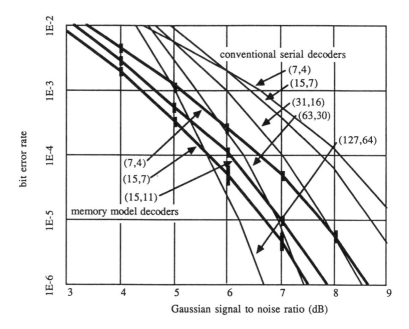

Figure 4.1. The neural network decoder applied to (7,4), (15,7), and (15,11) codes compared with conventional decoding performance quoted from [Gi]. The number pairs are $(n, \log_2 M)$. The simulation results are shown with 95% confidence intervals. The simulations were carried out by Peter W. Protzel, ICASE/NASA Langley.

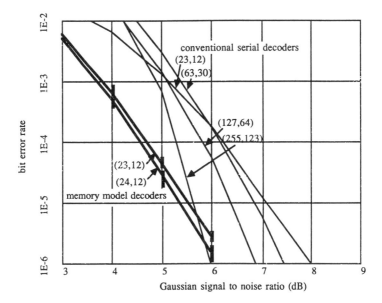

Figure 4.2. The neural network decoder applied to (23,12) and (24,12) codes compared with conventional decoding performance quoted from [Gi] and [ML]. The simulation results are shown with 95% confidence intervals. The simulations were carried out by Peter W. Protzel, ICASE/NASA Langley.

Figures 4.1 and 4.2 show that for a given signal to noise ratio, the error rate of large codes with conventional decoding can be achieved with much smaller codes and neural network decoding. For example, we see from Fig. 4.2 that with a signal to noise ratio of 6 dB, the error rate of conventional decoding of the (255, 123) code (with n = 255 and M = $2^{123} \cong$ 1E37 code words, an immense code) is about 1E-6. Approximately the same error rate can be achieved with a much smaller (24,12) code and the memory model.

4.3 General Recognition

Suppose we agree that the general observed state of some system can be expressed as a combination of numerical values. Suppose we wish to recognize patterns within the observed state of the system. Raw information from sensors, then, might consist of a long list of real numbers or a binary expression containing many more than 1000 bits. However, in terms of distinct memories, even 2^{1000} is for all practical purposes infinity. The task of recognition is to map raw information to a distinct memory or concept .

The memory model is a way to do this, that is, to store any M binary strings, $1 \leq M \leq 2^n$, with n artificial neurons. The input becomes a point in n-space (each component of which is a real number possibly expressed with many significant decimal places). This input becomes the initial state of a dynamical system which converges to one of the relatively few possible memories. Neuron output gain functions must be specified in a certain way (slopes must be greater than a finite number determined by n and M, but not infinite as with a step function). After such specification, it can be mathematically guaranteed that no spurious attractors exist, in contrast to other artificial neural network methods. If the number of accessible distinct memories is limited, then it is of course critical to preprocess the image from sensors in order to maximize the repertoire of recognizable inputs.

The problem of error correction of binary code is formally just a special case of general recognition. As an example of recognition rather different from code problems, the ten digits might be depicted in a rectangular array of pixels. In simulations of the memory model performed by S. Ruimveld [Ru] with the ten digits depicted in 4x5 pixel array (so M = 10 and n = 20, a very sparse code), convergence from a Hamming distance of 5 to the nearest memory has occurred typically in 30 time steps using $\varepsilon = 1E\text{-}6$ and $\Delta t = .1$. With a 10x12 pixel array convergence takes 350 time steps.

R. Carlson [CJ] has reduced the number of steps to convergence for the 10x12 array to about 50 by jumping in one first step from the initial state to an associated point on the boundary of the transistion zone; the associated point is the unique point lying in the boundary of T_ε and on the ray from the origin through the initial point.

A test of the memory model applied to 10x12 pixel recognition can be described as follows. To a memory with voltages -1 and +1 we add noise with noise voltages uniformly distributed between -1.5 and +1.5. Using simulation experience, ramp gain functions with $\varepsilon = 1E\text{-}38$ are chosen. A black pixel corresponds to $x_i \geq 1$, a white pixel to $x_i \leq -1$, and grey pixels to intermediate x_i values. The model will reliably recover to the original signal in about 50 time steps. A typical convergence sequence is shown in the following figures.

Fig. 4.3. The five before adding noise.

Fig. 4.4. A corrupted image obtained from the five by adding several doses of uniform noise from [-1.1, 1.1]. The corresponding point in 120-dimensional space is then fed to a memory model which has the ten digits stored as constant attractors in 120-dimensional space.

Fig. 4.5. The image after 37 iterations of the memory model.

Fig. 4.6. The image after 51 iterations and just prior to convergence.

Fig. 4.7. The five recognized by the memory model in 52 time steps.

4.4 Scanning in Image Recognition

Suppose now that we wish to detect one of four 4x5 pixel images (the digits 0, 1, 2, 3) somewhere within a 20x20 pixel array. There are 272 possible positions for the 4x5 image. Since it is impractical to use 272 copies of one memory, we can design a recognizer by scanning the 272 positions. At each scan position, the observed 20 pixel values are used as the initial state in 20-dimensional space for the memory model. A time limit of ten iterations is then imposed. If the system converges to one of the four memories (enters one of the four of 2^{20} orthants relative to the T_e), then the scanner is stopped. Otherwise the scanner moves to the next 4x5 array. R. Carlson has applied scanning to this problem with Gaussian noise. Typical results are shown in the following figures.

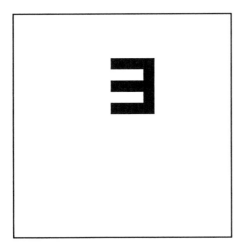

Fig. 4.8. The 4x5 three image place randomly in a 20x20 pixel array.

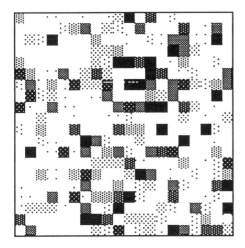

Fig. 4.9. The image after corruption with Gaussian noise voltages with a signal to noise ratio of 5.

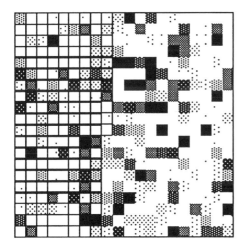

Fig. 4.10. The image after partial scanning.

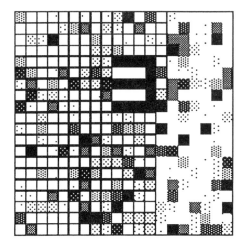

Fig. 4.11. The 4x5 three image found by scanning.

4.5 Commercial Neural Network Decoding

In 1986, J. C. Platt and J. J. Hopfield [PH] proposed error-correction of binary code using a neural network with system functions linear in g components, now known as the Hopfield model (see §3 of Chapter 1, especially p. 21, and §6 of Chapter 3). A code studied in their paper, called a *Viterbi code*, uses mxm binary matrices, so $n = m^2$. The code words are the permutation matrices. As m becomes large, this code becomes very sparse, that is, the ratio $m!(2^{m \cdot m})^{-1}$ of code words to possible words becomes a very small number. For m = 4 this fraction is $24(65536)^{-1}$ and for m = 10 it is only about 2.9E-24. Nonetheless, some Viterbi codes have been used in space communications with high levels of background noise.

It seems possible that the high order memory model might have a commercial future. It would be of particular interest to exploit the parallel nature of the memory model, perhaps by coupling electronic neurons directly to artificial retinal rods (pixel sensors arranged in an artificial retina) to avoid serial digital input. However, an essential element of this development will be the accurate multiplier implementing the image products. It is possible to speculate that optical chips would be well suited to this task, but the rate of progress in optical chip technology seems difficult to estimate.

Lastly, we mention that it is possible by completely different methods to use a type of composition function neural network to solve the error-correcting code problem. Using this approach a system of functions is *trained*, that is, *weights* are determined by a method relying upon many examples of the conversion from code word with error to intended code word. A successful application of this method is in [HSB].

PROBLEMS

1. Devise a spreadsheet which models the memory model decoder for the Hamming (7,4) code in Table 4.1.

2. Devise a spreadsheet which models the memory model decoder for the Hadamard code in Table 4.2.

ANSWERS

Spreadsheets for the memory models are shown on the following pages.

	A	B	C	D	E	F	G	H
1	delta t =	0.375		delta =		epsilon =		
2	rand_start	0.08525		0		0.1		
3								
4	m1 =	0	0	0	0	0	0	0
5	m2 =	0	0	0	1	1	1	1
6	m3 =	0	0	1	0	1	1	0
7	m4 =	0	0	1	1	0	0	1
8	m5 =	0	1	0	0	1	0	1
9	m6 =	0	1	0	1	0	1	0
10	m7 =	0	1	1	0	0	1	1
11	m8 =	0	1	1	1	1	0	0
12	m9 =	1	0	0	0	0	1	1
13	m10 =	1	0	0	1	1	0	0
14	m11 =	1	0	1	0	1	0	1
15	m12 =	1	0	1	1	0	1	0
16	m13 =	1	1	0	0	1	1	0
17	m14 =	1	1	0	1	0	0	1
18	m15=	1	1	1	0	0	0	0
19	m16=	1	1	1	1	1	1	1
20	input =	0.9	-0.8	1	1.1	-1	-0.7	-0.9
21		0.9	0.8	1	1.1	1	0.7	0.9
22	0.7	0.12857	-0.11429	0.14286	0.15714	-0.14286	-0.1	-0.12857
23	start_cell	0						
24	x =	0.01837	-0.01633	0.02041	0.02245	-0.02041	-0.01429	-0.01837
25	g(x) =	0.59184	0.41837	0.60204	0.61224	0.39796	0.42857	0.40816
26	ramp g =	1	ramp jump	1	sigmoid g	1		
27	m1i*{gi} =	0.40816	0.58163	0.39796	0.38776	0.60204	0.57143	0.59184
28	m2i*{gi} =	0.40816	0.58163	0.39796	0.61224	0.39796	0.42857	0.40816
29	m3i*{gi} =	0.40816	0.58163	0.60204	0.38776	0.39796	0.42857	0.59184
30	m4i*{gi} =	0.40816	0.58163	0.60204	0.61224	0.60204	0.57143	0.40816
31	m5i*{gi} =	0.40816	0.41837	0.39796	0.38776	0.39796	0.57143	0.40816
32	m6i*{gi} =	0.40816	0.41837	0.39796	0.61224	0.60204	0.42857	0.59184
33	m7i*{gi} =	0.40816	0.41837	0.60204	0.38776	0.60204	0.42857	0.40816
34	m8i*{gi} =	0.40816	0.41837	0.60204	0.61224	0.39796	0.57143	0.59184
35	m9i*{gi} =	0.59184	0.58163	0.39796	0.38776	0.60204	0.42857	0.40816
36	m10i*{gi}	0.59184	0.58163	0.39796	0.61224	0.39796	0.57143	0.59184
37	m11i*{gi}	0.59184	0.58163	0.60204	0.38776	0.39796	0.57143	0.40816
38	m12i*{gi}	0.59184	0.58163	0.60204	0.61224	0.60204	0.42857	0.59184
39	m13i*{gi}	0.59184	0.41837	0.39796	0.38776	0.39796	0.42857	0.59184
40	m14i*{gi}	0.59184	0.41837	0.39796	0.61224	0.60204	0.57143	0.40816
41	m15i*{gi}	0.59184	0.41837	0.60204	0.38776	0.60204	0.57143	0.59184
42	m16i*{gi}	0.59184	0.41837	0.60204	0.61224	0.39796	0.42857	0.40816
43								
44	I1 =	0.00746						
45	I2 =	0.00403						
46	I3 =	0.00559						
47	I4 =	0.01229						
48	I5 =	0.00245						
49	I6 =	0.00635						
50	I7 =	0.0042						
51	I8 =	0.00847						
52	I9 =	0.00559						

	A	B	C	D	E	F	G	H
53	I10 =	0.01129						
54	I11 =	0.00746						
55	I12 =	0.01938						
56	I13 =	0.00386						
57	I14=	0.00847						
58	I15 =	0.01177						
59	I16 =	0.00635						
60					0.01938			
61	(2mij-1)rj	-0.00746	-0.00746	-0.00746	-0.00746	-0.00746	-0.00746	-0.00746
62		-0.00403	-0.00403	-0.00403	0.00403	0.00403	0.00403	0.00403
63		-0.00559	-0.00559	0.00559	-0.00559	0.00559	0.00559	-0.00559
64		-0.01229	-0.01229	0.01229	0.01229	-0.01229	-0.01229	0.01229
65		-0.00245	0.00245	-0.00245	-0.00245	0.00245	-0.00245	0.00245
66		-0.00635	0.00635	-0.00635	0.00635	-0.00635	0.00635	-0.00635
67		-0.0042	0.0042	0.0042	-0.0042	-0.0042	0.0042	0.0042
68		-0.00847	0.00847	0.00847	0.00847	0.00847	-0.00847	-0.00847
69		0.00559	-0.00559	-0.00559	-0.00559	-0.00559	0.00559	0.00559
70		0.01129	-0.01129	-0.01129	0.01129	0.01129	-0.01129	-0.01129
71		0.00746	-0.00746	0.00746	-0.00746	0.00746	-0.00746	0.00746
72		0.01938	-0.01938	0.01938	0.01938	-0.01938	0.01938	-0.01938
73		0.00386	0.00386	-0.00386	-0.00386	0.00386	0.00386	-0.00386
74		0.00847	0.00847	-0.00847	0.00847	-0.00847	-0.00847	0.00847
75		0.01177	0.01177	0.01177	-0.01177	-0.01177	-0.01177	-0.01177
76		0.00635	0.00635	0.00635	0.00635	0.00635	0.00635	0.00635
77								
78	next x =	0.02023	-0.01814	0.02251	0.02462	-0.02251	-0.01429	-0.02023
79								
80	loop =	0.02023	-0.01814	0.02251	0.02462	-0.02251	-0.01429	-0.02023

	A	B	C	D	E	F	G	H	I	J	K	L
1	delta t	0.5		delta =		epsilon =						
2	rand st	0.618		0		2E-04						
3	m1 =	0	0	0	0	0	0	0	0	0	0	0
4	m2 =	1	1	0	1	1	1	0	0	0	1	0
5	m3 =	0	1	1	0	1	1	1	0	0	0	1
6	m4 =	1	0	1	1	0	1	1	1	0	0	0
7	m5 =	0	1	0	1	1	0	1	1	1	0	0
8	m6 =	0	0	1	0	1	1	0	1	1	1	0
9	m7 =	0	0	0	1	0	1	1	0	1	1	1
10	m8 =	1	0	0	0	1	0	1	1	0	1	1
11	m9 =	1	1	0	0	0	1	0	1	1	0	1
12	m10 =	1	1	1	0	0	0	1	0	1	1	0
13	m11 =	0	1	1	1	0	0	0	1	0	1	1
14	m12 =	1	0	1	1	1	0	0	0	1	0	1
15	m13 =	0	0	1	0	0	0	1	1	1	0	1
16	m14 =	1	0	0	1	0	0	1	1	1	1	0
17	m15=	0	1	0	0	1	0	0	0	1	1	1
18	m16=	1	0	1	0	0	1	0	0	0	1	1
19	m17=	1	1	0	1	0	0	1	0	0	0	1
20	m18=	1	1	1	0	1	0	0	1	0	0	0
21	m19=	0	1	1	1	0	1	0	0	1	0	0
22	m20=	0	0	1	1	1	0	1	0	0	1	0
23	m21=	0	0	0	1	1	1	0	1	0	0	1
24	m22=	1	0	0	0	1	1	1	0	1	0	0
25	m23=	0	1	0	0	0	1	1	1	0	1	0
26	m24=	1	1	1	1	1	1	1	1	1	1	1
27												
28	start c	0										
29	x =	1	1	1	- 1	- 1	- 1	1	- 1	1	1	- 1
30	g(x) =	1	1	1	0	0	0	1	0	1	1	0
31	sigmoid	1	ramp g	1	jump ra	0.5						
32	m1i*{gi	0	0	0	1	1	1	0	1	0	0	1
33	m2i*{gi	1	1	0	0	1	0	0	1	0	1	1
34	m3i*{gi	0	1	1	1	0	0	1	1	0	0	0
35	m4i*{gi	1	0	1	0	1	0	1	0	0	0	1
36	m5i*{gi	0	1	0	0	0	1	1	0	1	0	1
37	m6i*{gi	0	0	1	1	0	0	0	0	1	1	1
38	m7i*{gi	0	0	0	0	1	0	1	1	1	1	0
39	m8i*{gi	1	0	0	1	0	1	1	0	0	1	0
40	m9i*{gi	1	1	0	1	1	0	0	0	1	0	0
41	m10i*{	1	1	1	1	1	1	1	1	1	1	1
42	m11i*{	0	1	1	0	1	1	0	0	0	1	0
43	m12i*{	1	0	1	0	0	1	0	1	1	0	0
44	m13i*{	0	0	1	1	1	1	1	0	1	0	0
45	m14i*{	1	0	0	0	1	1	0	0	1	1	1
46	m15i*{	0	1	0	1	0	1	0	1	1	1	0
47	m16i*{	1	0	1	1	1	0	0	1	0	1	0
48	m17i*{	1	1	0	0	1	1	1	1	0	0	0
49	m18i*{	1	1	1	1	0	1	0	0	0	0	1
50	m19i*{	0	1	1	0	1	0	0	1	1	0	1
51	m20i*{	0	0	1	0	0	1	1	1	0	1	1
52	m21i*{	0	0	0	0	0	0	0	0	0	0	0

	A	B	C	D	E	F	G	H	I	J	K	L
53	m22i*{	1	0	0	1	0	0	1	1	1	0	1
54	m23i*{	0	1	0	1	1	0	1	0	0	1	1
55	m24i*{	1	1	1	0	0	0	1	0	1	1	0
56												
57	I1 =	0										
58	I2 =	0										
59	I3 =	0										
60	I4 =	0										
61	I5 =	0										
62	I6 =	0										
63	I7 =	0										
64	I8 =	0										
65	I9 =	0										
66	I10 =	1										
67	I11 =	0										
68	I12 =	0										
69	I13 =	0										
70	I14=	0										
71	I15 =	0										
72	I16 =	0										
73	I17 =	0										
74	I18 =	0										
75	I19 =	0										
76	I20 =	0										
77	I21 =	0										
78	I22 =	0										
79	I23 =	0										
80	I24 =	0										
81													
82	(2mij-	0	0	0	0	0	0	0	0	0	0	0
83		0	0	0	0	0	0	0	0	0	0	0
84		0	0	0	0	0	0	0	0	0	0	0
85		0	0	0	0	0	0	0	0	0	0	0
86		0	0	0	0	0	0	0	0	0	0	0
87		0	0	0	0	0	0	0	0	0	0	0
88		0	0	0	0	0	0	0	0	0	0	0
89		0	0	0	0	0	0	0	0	0	0	0
90		0	0	0	0	0	0	0	0	0	0	0
91		1	1	1	-1	-1	-1	1	-1	1	1	-1
92		0	0	0	0	0	0	0	0	0	0	0
93		0	0	0	0	0	0	0	0	0	0	0
94		0	0	0	0	0	0	0	0	0	0	0
95		0	0	0	0	0	0	0	0	0	0	0
96		0	0	0	0	0	0	0	0	0	0	0
97		0	0	0	0	0	0	0	0	0	0	0
98		0	0	0	0	0	0	0	0	0	0	0
99		0	0	0	0	0	0	0	0	0	0	0
100		0	0	0	0	0	0	0	0	0	0	0
101		0	0	0	0	0	0	0	0	0	0	0
102		0	0	0	0	0	0	0	0	0	0	0
103		0	0	0	0	0	0	0	0	0	0	0
104		0	0	0	0	0	0	0	0	0	0	0

	A	B	C	D	E	F	G	H	I	J	K	L
105		0	0	0	0	0	0	0	0	0	0	0
106												
107	next x	1	1	1	- 1	- 1	- 1	1	- 1	1	1	- 1
108												
109	loop =	1	1	1	- 1	- 1	- 1	1	- 1	1	1	- 1

Chapter 5--Memory Models with Limit Cycles as Attractors

5.1 A Two-Dimensional Limit Cycle

In this chapter our goal is to design a version of (1.1), specifically a modification of the memory model (3.5), which offers dense storage of limit cycles as attractors.

Let us briefly recall model (1.3). The dynamics of two neurons are given by

$$dx_1/dt = -x_1 - g_1g_2 - 2(1-g_1)g_2 + (1-g_1)(1-g_2) + 2g_1(1-g_2) \tag{1.3}$$
$$dx_2/dt = -x_2 + 2g_1g_2 - (1-g_1)g_2 - 2(1-g_1)(1-g_2) + g_1(1-g_2)$$

where the gain function g is a ramp function with $\varepsilon = \varepsilon_1 = \varepsilon_2$. The linear approximation matrix for (1.3) at the only constant trajectory (0,0) is

$$L = \begin{pmatrix} -1+g' & -3g' \\ 3g' & -1+g' \end{pmatrix}$$

where $g' = 1/(2\varepsilon)$. Therefore, (0,0) is unstable if $g' > 1$, that is, $.5 > \varepsilon$. Representative trajectories of (1.3) with $\varepsilon = .1$ including the limit cycle are as shown in Fig. 1.2 .

In the generalization of (1.3), we will be able to store multiple limit cycles and constant attractors as memories. With each limit cycle and its tour of certain orthants relative to T_ε, we can associate a *cyclic list of g values* such as, for example, $\rightarrow(1,1)\rightarrow(0,1)\rightarrow(0,0)\rightarrow(1,0)\rightarrow(1,1)\rightarrow$ in (1.3). We shall only require that consecutive values for g in cyclic lists change in only one component and that the various limit cycles and constant trajectories have no g values in common.

Note that if p is the number of distinct g values in a cyclic list, then p is even and at least 4. This is so because only one component changes in consecutive values for g and for every change in component i there must exist

102

somewhere in the list the reverse change.

From Chapter 3, a set of (constant) *memories* is a set of binomial strings $\{m_\iota\}$, $\iota = 1,..., M$, having components $m_{\iota j} = 1$ or 0, $j = 1,...,n$. We say m_ι is *stored* as a constant trajectory $\underline{x}$ for (1.1) if $g_j(x_j) = m_{\iota j}$. From given memories $\{m_\iota\}$ we can form *image products* $\{I_\iota\}$ as

$$I_\iota = \prod_{j=1}^{n} \{m_{\iota j} g_j(x_j) + (1 - m_{\iota j})(1 - g_j(x_j))\} \qquad (3.1)$$

For example, if n = 10 and $m_1 = (0,1,1,1,0,0,1,0,1,1)$, then $I_1 = (1-g_1)g_2g_3g_4(1-g_5)(1-g_6)g_7(1-g_8)g_9g_{10}$.

The *memory model* which stores memories $\{m_\iota\}$ as attractors is

$$dx_i/dt = -x_i + \sum_{\iota=1}^{M} (2m_{\iota i} - 1) r_\iota (I_\iota (g)) \qquad (3.5)$$

We proceed to devise a dynamical system with the property that any trajectory which starts in an orthant relative to T_ε with g values in a cyclic list, converges to prespecified simple closed curve in n-space, a limit cycle. Any trajectory which starts exactly on the curve, stays on the curve and is cyclic with some finite period. Furthermore, the g values observed along any such trajectory are those in the cyclic list (except as the trajectory momentarily passes through T_ε from one orthant to another). A neural network model with this feature might be thought of as having memorized a cyclic signal or waveform.

By identifying (0,1)-strings and (-1,1)-strings, each (0,1) n-string represents an orthant of n-space. Let $\{m_\iota\}$ be a cyclic list, $\iota = 1, 2, ..., p$. For algebraic purposes define $m_0 \equiv m_p$, $m_{p+1} \equiv m_1$, and $1 \equiv (1,1,...,1)$. Thus the p points $\{ ..., (m_{\iota-1}+m_\iota-1), (m_\iota+m_{\iota+1}-1),... \}$, $\iota = 1, 2, ..., p$, can be identified with p points in n-dimensional space with vertex coordinates 0 , $+1$, or -1. Consecutive points differ in exactly two components and the difference in those components is 1. Restricting attention to precisely the two components that do change in a consecutive pair of the points, the change must be between some pair $(\pm1,0)$ and $(0,\pm1)$. In the language of graph theory, the edges joining consecutive vertices in the list comprise a *p-cycle*.

Suppose we wish to associate with the p-cycle a limit cycle that tours orthants relative to T_ϵ in the order of the vertices. As in the memory model, m_{ti} denotes the ith entry in m_t. Consider the following modification of the memory model:

$$dx_i/dt = -x_i + \sum_{t=1}^{M} (-m_{t-1i}+m_{ti}+2m_{t+1i}-1)I_t \ (g) \qquad (5.1)$$

We next establish that this model stores the given cycle as a limit cycle.

THEOREM 4.1. For sufficiently small ϵ, the model (5.1) has an attractor trajectory, namely a limit cycle which tours the orthants of n-space in the order of the p-cycle.

PROOF. Let us modify the concept of gain function to include a limiting case of $\epsilon \to 0$, that is, the step function $g_i = g_i(x_i)$ defined by $g_i = 0$ if $x_i < 0$; $g_i = 1$ if $x_i > 0$; and $g_i (0) = .5$. Consider the trajectory which starts at a point on the boundary which separates the orthants containing m_0 and m_1, namely at the point m_0+m_1-1. This trajectory traverses the orthant containing m_1 (where $I_1 = 1$ and all other $I_i = 0$) according to the equations

$$dx_i/dt = -x_i - m_{0i}+ m_{1i}+2m_{2i} -1$$

that is, the trajectory follows the straight line connecting $m_0+ m_1-1$ and $m_1+ m_2-1$ until $m_1+ m_2-1$ is reached. Explicitly the trajectory is

$$x_i = (- m_{0i}+ m_{1i}+2m_{2i} -1) +2(m_0-m_2)e^{-t}$$

for $0 \le t \le ln\ 2$. This procedure extends to other orthants associated with other entries in the p-cycle until a cyclic trajectory made of p linear pieces in the p orthants is obtained.

We note that in each such orthant (where all trajectories are converging straight lines) that all other trajectories approach the given trajectory. Thus if a trajectory of (5.1) with $\epsilon = 0$ starts in any orthant containing some m_t, it must thereafter asymptotically approach the given cyclic trajectory. On the other hand, a trajectory which starts in an orthant containing no m_t satisfies $dx_i/dt = -x_i$, that is, consists of a straight line trajectory which asymptotically approaches the origin.

Now we return to gain functions with finite slopes at 0 and the associated transition zone T_ε. For given ε_i, we can produce a worst estimate of how much a new trajectory starting at $.5m_0+m_1+.5\ m_2-1$ deviates from the ε_i $= 0$ cyclic trajectory; then we can choose ε sufficiently small so that the new trajectory tours the same orthants and is still stable in each orthant. It follows that the new trajectory is also a stable limit cycle. Furthermore, a trajectory starting in an orthant containing no m_t now approaches the origin only until it enters T_ε after a finite time interval.   QED

Simulations have indicated that ramp gain functions with $\varepsilon > 0$ and various other types of gain functions can be used in (4.1), providing ε is sufficiently small.

As an example, suppose we wish to generate a model with a limit cycle which tours the 6 memories

$$\rightarrow(1,1,1)\rightarrow(1,1,0)\rightarrow(0,1,0)\rightarrow(0,1,1)\rightarrow(0,0,1)\rightarrow(1,0,1)\rightarrow(1,1,1)\rightarrow$$

That is, we use

$$
\begin{aligned}
m_0 &= (1,0,1) \\
m_1 &= (1,1,1) \\
m_2 &= (1,1,0) \\
m_3 &= (0,1,0) \\
m_4 &= (0,1,1) \\
m_5 &= (0,0,1) \\
m_6 &= (1,0,1) \\
m_7 &= (1,1,1)
\end{aligned}
\qquad (5.2)
$$

where m_0 and m_7 are defined as above as m_6 and m_1. We can think of the limit cycle in terms of orthants as the set of edges in Fig. 5.1.

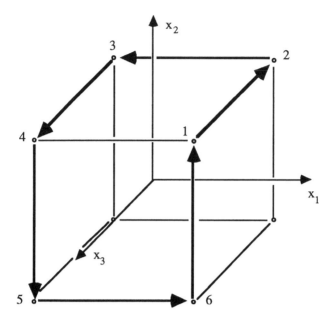

Figure 5.1. A tour of six of the eight orthants of three-dimensional space represented as edges of a cube with vertex components ±1 .

However, the trajectory of the system touring these orthants using the model (5.1) actually looks like that shown in Fig. 5.2.

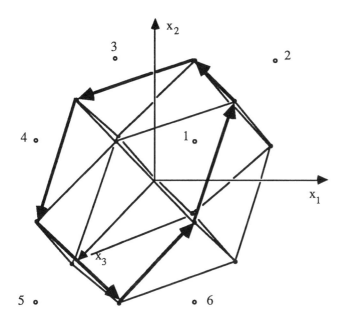

Figure 5.2. The trajectory segments for a tour of six orthants represented as edges of a cuboctahedron. Note that the swerving of the trajectory through the transition zone is not shown at the arrow points.

We note that any limit cycle for a three-dimensional model based on (5.1) and (5.2) must proceed along the edges of the cuboctahedron in Fig. 5.2 using at most one edge in each of the triangular faces.

5.2 Wiring

Let us compare the two-neuron memory model storing all four of the possible memories

$$dx_1/dt = -x_1 + g_1g_2 + g_1(1-g_2) - (1-g_1)g_2 - (1-g_1)(1-g_2) \qquad (5.3)$$
$$dx_2/dt = -x_2 + g_1g_2 - g_1(1-g_2) + (1-g_1)g_2 - (1-g_1)(1-g_2)$$

and the two-neuron memory model storing a limit cycle

$$dx_1/dt = -x_1 -g_1g_2 -2(1-g_1)g_2+(1-g_1)(1-g_2)+2g_1(1-g_2)$$ (1.3)
$$dx_2/dt = -x_2 +2g_1g_2 -(1-g_1)g_2-2(1-g_1)(1-g_2)+g_1(1-g_2)$$

The dynamics of these systems are shown in Fig. 5.3 .

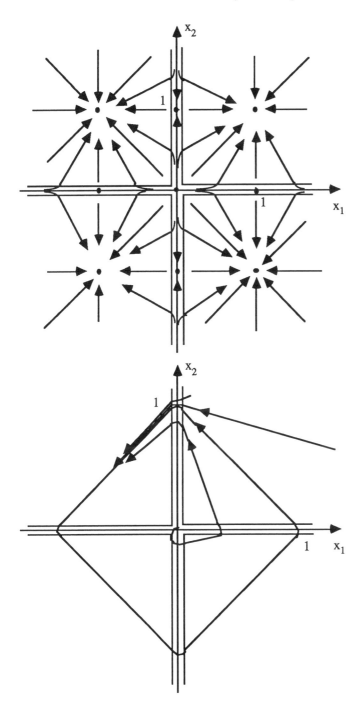

Figure 5.3. Contrast of dynamics of systems (5.3) and (1.3).

The schematic of system (5.3) is shown in Fig. 5.4 .

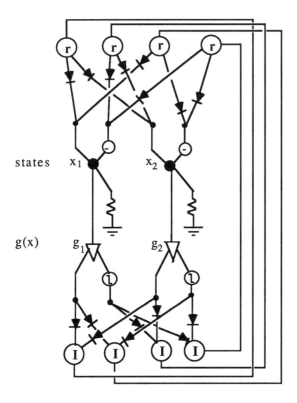

Figure 5.4. The schematic circuit corresponding to a two-neuron system which stores the four constant memories (1,1), (1,0), (0,1), and (0,0).

Note the matching between neuron output connections and input connections. An analogous schematic for the limit cycle model (1.3) would entail permuting and weighting the connections as shown in Fig. 5.5.

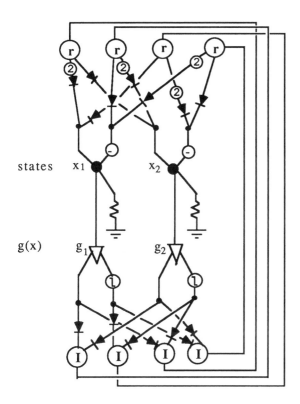

Figure 5.5. The schematic circuit corrseponding to system (1.3).

5.3 Neural Networks with a Mixture of Limit Cycles and Constant Attractors

Suppose q disjoint p-cycles (of various lengths) are specified, perhaps together with r other binomial n-strings not listed in any of the cycles. Specifying the nonlinear part of the right side of (1.1) to be a sum of terms from (3.3) and (5.1) (with all $k_i = 1$) yields a model with q associated limit cycles and r constant attractors.

As an example, we might store with three neurons the cycle list used above in (5.2)

$$\to(1,1,1)\to(1,1,0)\to(0,1,0)\to(0,1,1)\to(0,0,1)\to(1,0,1)\to(1,1,1)\to$$

together with the constant memory (0,0,0). The required image products are for the limit cycle: $I_1 = g_1g_2g_3$; $I_2 = g_1g_2(1-g_3)$; $I_3 = (1-g_1)g_2(1-g_3)$; $I_4 = (1-g_1)g_2g_3$; $I_5 = (1-g_1)(1-g_2)g_3$; $I_6 = g_1(1-g_2)g_3$; and for the constant trajectory $I_7 = (1-g_1)(1-g_2)(1-g_3)$. Using the above recipe leads to the dynamical system

$$dx_1/dt = -x_1 + I_1 \quad - I_2 \ -2I_3 -I_4 \quad +I_5 \quad +2I_6 \quad -I_7$$
$$dx_2/dt = -x_2 + 2I_1 +I_2 \ +I_3 \ -I_4 \quad -2I_5 \ +I_6 \quad -I_7$$
$$dx_3/dt = -x_3 - I_1 - 2I_2 \ +I_3 \ +2I_4 +I_5 \ +I_6 \quad -I_7$$

PROBLEMS

1. Design a three-dimensional memory model with a limit cycle which tours all eight orthants.

2. Design a five-dimensional memory model with two limit cycles touring orthants associated with:

$\rightarrow(1,1,1,1,1)\rightarrow(1,1,1,1,0)\rightarrow(1,1,1,0,0)\rightarrow(1,1,1,0,1)\rightarrow(1,1,1,1,1)\rightarrow$

$\rightarrow(0,1,1,1,1)\rightarrow(0,0,1,1,1)\rightarrow(0,0,0,1,1)\rightarrow(0,0,0,0,1)\rightarrow(0,1,0,0,1)\rightarrow$
$\rightarrow(0,1,1,0,1)\rightarrow(0,1,1,1,1)\rightarrow$

together with the constant memories (0,0,0,0,0,) and (1,0,0,0,0).

ANSWERS

1. One tour of all eight orthants is

$$\rightarrow(1,1,1)\rightarrow(1,1,0)\rightarrow(1,0,0)\rightarrow(1,0,1)\rightarrow(0,0,1)\rightarrow(0,0,0)\rightarrow(0,1,0)\rightarrow\rightarrow(0,1,1)\rightarrow$$
$$\rightarrow(1,1,1)\rightarrow$$

The required image products are

$I_1 = g_1 g_2 g_3$; $I_2 = g_1 g_2 (1-g_3)$; $I_3 = g_1 (1-g_2)(1-g_3)$; $I_4 = g_1 (1-g_2) g_3$;
$I_5 = (1-g_1)(1-g_2) g_3$; $I_6 = (1-g_1)(1-g_2)(1-g_3)$; $I_7 = (1-g_1) g_2 (1-g_3)$;
$I_8 = (1-g_1) g_2 g_3$

Using (5.1) leads to the dynamical system

$dx_1/dt = -x_1 + 2I_1 + I_2 + I_3 - I_4 - 2I_5 - I_6 - I_7 + I_8$
$dx_2/dt = -x_2 + I_1 - I_2 - 2I_3 - I_4 - I_5 + I_6 + 2I_7 + I_8$
$dx_3/dt = -x_3 - I_1 - 2I_2 + I_3 + 2I_4 - I_5 - 2I_6 + I_7 + 2I_8$

2. The required image products for the first limit cycle are

$I_1 = g_1 g_2 g_3 g_4 g_5$; $I_2 = g_1 g_2 g_3 g_4 (1-g_5)$; $I_3 = g_1 g_2 g_3 (1-g_4)(1-g_5)$;
$I_4 = g_1 g_2 g_3 (1-g_4) g_5$

The required image products for the second limit cycle are

$I_5 = (1-g_1) g_2 g_3 g_4 g_5$; $I_6 = (1-g_1)(1-g_2) g_3 g_4 g_5$; $I_7 = (1-g_1)(1-g_2)(1-g_3) g_4 g_5$;
$I_8 = (1-g_1)(1-g_2)(1-g_3)(1-g_4) g_5$; $I_9 = (1-g_1) g_2 (1-g_3)(1-g_4) g_5$;
$I_{10} = (1-g_1) g_2 g_3 (1-g_4) g_5$;

The required image products for the constant trajectories are

$I_{11} = (1-g_1)(1-g_2)(1-g_3)(1-g_4)(1-g_5)$; $I_{12} = g_1 (1-g_2)(1-g_3)(1-g_4)(1-g_5)$;

This leads to the memory model (5.1) with specified dynamics:

$dx_1/dt = -x_1 + I_1 + I_2 + I_3 + I_4 - I_5 - I_6 - I_7 - I_8 - I_9 - I_{10} - I_{11} + I_{12}$

$dx_2/dt = -x_2 + I_1 + I_2 + I_3 + I_4 - I_5 - 2I_6 - I_7 + I_8 + 2I_9 + I_{10} - I_{11} - I_{12}$

$dx_3/dt = -x_3 + I_1 + I_2 + I_3 + I_4 + I_5 - I_6 - 2I_7 - I_8 + I_9 + 2I_{10} - I_{11} - I_{12}$

$dx_4/dt = -x_4 + 2I_1 - I_2 - 2I_3 + I_4 + 2I_5 + I_6 - I_7 - 2I_8 - I_9 + I_{10} - I_{11} - I_{12}$

$dx_5/dt = -x_5 - I_1 - 2I_2 + I_3 + 2I_4 + I_5 + I_6 + I_7 + I_8 + I_9 + I_{10} - I_{11} - I_{12}$

Spreadsheets for generating trajectories for the systems in these two problems are shown on the following pages.

	A	B	C	D	E	F	G	H	I	J
1	delta t =	0.1	epsilon =	0.01						
2										
3	start cell	0	random =	0.2444						
4	x =	0.18922	g(x) =	1	1-g(x) =	0				
5	y =	-1.036	g(y) =	0	1-g(y) =	1				
6	z =	0.8299	g(z) =	1	1-g(z) =	0				
7		I1 =	I2 =	I3 =	I4 =	I5 =	I6 =	I7 =	I8 =	
8		0	0	0	1	0	0	0	0	0
9										
10	0	1	1	1	1	0	0	0	0	1
11	1	1	1	0	0	0	0	1	1	1
12	1	1	0	0	1	1	0	0	1	1
13										
14		2	1	1	-1	-2	-1	-1	1	
15		1	-1	-2	-1	-1	1	2	1	
16		-1	-2	1	2	-1	-2	1	2	
17										
18	next x =	0.0703								
19	next y =	-1.0324								
20	next z =	0.94691								
21										
22	loop	0.0703								
23		-1.0324								
24		0.94691								

	A	B	C	D	E	F	G	H
1	delta t =	0.1	epsilon =	0.01				
2								
3	start cell =	0	random =	-0.17159				
4	x1=	1	g1=	1	1-g1=	0		
5	x2=	1	g2=	1	1-g2=	0		
6	x3=	1	g3=	1	1-g3=	0		
7	x4=	-0.42231	g4=	0	1-g4=	1		
8	x5=	0.731641	g5=	1	1-g5=	0		
9								
10		I1 =	I2 =	I3 =	I4 =			
11	1	0	0	0	1	0		
12								
13		I5 =	I6 =	I7 =	I8 =	I9=	I10=	
14	0	0	0	0	0	0	0	0
15								
16		I11=	I12=					
17		0	0					
18								
19	next x1=	1						
20	next x2=	1						
21	next x3=	1						
22	next x4=	-0.28008						
23	next x5=	0.858477						
24								
25	loop	1						
26		1						
27		1						
28		-0.28008						
29		0.858477						

Chapter 6--Solving Operations Research Problems with Neural Networks

6.1 Selecting Permutation Matrices with Neural Networks

As stated in Chapter 1, neural network models are in principle capable of solving certain finite choice problems arising in operations research. Generally we wish to build a dynamical system which has as its attractors feasible solutions of a finite choice problem. Also, we wish to design the regions of attraction so that optimal or near optimal solutions are favored when the system is started at a random point x in n-space with $|x_i| < 1$. Thus the attractors (answers) are implicitly built into the system functions of the model and are discovered by repeated simulations starting at randomly selected points inside the n-cube with vertex components ± 1.

We shall focus on finite choice problems which are answered by m×m permutation matrices with $m \geq 2$. We can think of the answers as placements of m nonthreatening rooks on an m×m chessboard. Thus the dimension of the neural network is $n = m^2$ and any linear approximation matrix at a constant trajectory is $m^2 \times m^2$. For the sake of simplicity all ε_i are assumed equal, $\varepsilon_i = \varepsilon$.

We first consider the set selection model for choosing exactly one neuron on in each row and exactly one neuron on in each column. The general set selection model from Chapter 1

$$dx_i/dt = -x_i + 1 - (2/n_i) \sum_{\substack{S_j \\ i \in S_j}} \sum_{\substack{k \in S_j \\ k \neq i}} g_k(x_k) \tag{1.5}$$

becomes

$$dx_i/dt = -x_i + 1 - \sum_{\substack{j \in R_i \\ j \neq i}} g_j(x_j) - \sum_{\substack{j \in C_i \\ j \neq i}} g_j(x_j) \tag{6.1}$$

where R_i is row i and C_i is column i. For the sake of brevity we define $A_i \equiv R_i \cup C_i - \{i\}$ so (5.1) can be written

$$dx_i/dt = -x_i + 1 - \sum_{j \in A_i} g_j(x_j) \qquad (6.2)$$

Because we will incorporate quantitative data in subsequent optimization problems, we actually wish to study a generalization of (5.2) with positive benefit coefficients $\{b_i\}$ (inverse of cost coefficients) given by

$$dx_i/dt = -x_i + b_i \left\{ 1 - \sum_{j \in A_i} g_j(x_j) \right\} \qquad (6.3)$$

The proof of Theorem 2.4 guarantees that all trajectories of (6.2) to be or to approach asymptotically constant trajectories. Modifying that proof by specifying $\lambda_i = -b_i$ suffices to show the same for (6.3).

Clearly, answer sets (and associated constant trajectories with x components ± 1) for (6.2) correspond one-to-one with answer sets (and associated constant trajectories with x components $\pm b_i$) for (6.3) so long as $\varepsilon < .5b_i$. We assume $\varepsilon < .5b_i$ in all that follows.

Let us consider the general constant trajectories of (6.3), that is, the general solutions of

$$x_i = + b_i \left\{ 1 - \sum_{j \in A_i} g_j(x_j) \right\} \qquad (6.4)$$

Any answer set (permutation matrix) for (6.3) gives rise to other answer sets by permuting the rows or columns of the chessboard. This just amounts to relabeling the neurons and subsets. It follows that any constant trajectory for (6.3), even a constant trajectory with some g_i between 0 and 1, gives rise to other constant trajectories by such permutations.

We proceed to characterize the general constant trajectories of (6.3).

THEOREM 6.1. Suppose a constant trajectory x for (6.3) has every g = 0 or 1, that is, every $x_i < -\varepsilon_i$ or $x_i > \varepsilon_i$. Then exactly one neuron is on in each row and column.

PROOF. We follow the proof of Theorem 1.4. Suppose two neurons x_i and x_j are on in some row. Then (6.4) implies $x_i \leq 0$, a contradiction. Similarly at most one neuron can be on in each column. Suppose xi is off and all other neurons in the same row are off. Then $x_i < 0$. Then (6.4) implies that at least two neurons in the same column as x_i are on, a contradiction. Thus exactly one neuron is on in each row and column.   QED

As a corollary we see that if a constant trajectory x for (6.3) has every g = 0 or 1, that is, every $x_i < -\varepsilon_i$ or $x_i > \varepsilon_i$, then each x_i is $\pm b_i$.

If a constant trajectory has $x_i = 1$ or $x_i = 0$, we say neuron i is *on* or *off*. If a constant trajectory has $|x_i| < \varepsilon$ (so $0 < g_i < 1$), we say neuron i is *transitional*. We proceed to obtain information about constant trajectories of (6.3) with some transitional neurons.

THEOREM 6.2. (from the dissertation of G. Tagliarini [T]) Suppose a constant trajectory for (6.3) has one transitional neuron. Then at least one other neuron must be transitional.

PROOF. For the purpose of organizing our proof, let us label the m^2 squares of the chessboard starting in the upper left corner, left to right, top to bottom. Suppose the statement is false. Without loss of generality, we assume neuron 1 is in transition and all other neurons are on or off. In solving for x_1 in (6.4), we see that the term in { } is an integer of magnitude less than 1, that is, 0. Thus exactly one neuron in A_1 must be on and all others off. We may as well assume that neuron 2 is on (so all other neurons in row 1 and column 1 are off). Since x_2 must then be positive, the i = 2 instance of (6.4) implies all other neurons in the second column must be off as well.

Consider any neuron k in the first column, k ≠ 1. Since k is off, the kth instance of (6.4) implies at least one neuron in the row containing k must be on. That neuron is not k itself or the neuron in the second column. Repeating this argument for all the neurons in the first column yields m-1 neurons on somewhere in the positions below the first row 1 and to the right of the second column. This implies some such row or column has two neurons on, contradicting Theorem 6.1.   QED

THEOREM 6.3 (from the dissertation of G. Tagliarini [T]). Suppose in an ε-invariant constant trajectory for (6.3) neuron i is in transition. Then there must actually be at least one other transitional neuron in A_i.

PROOF. Suppose the statement is false. Without loss of generality, we assume all $s \geq 2$ neurons in transition are in the upper left sxs block positions of the chessboard shown in Fig. 6.1. In each row and each column of the sxs block exactly one neuron in is transition.

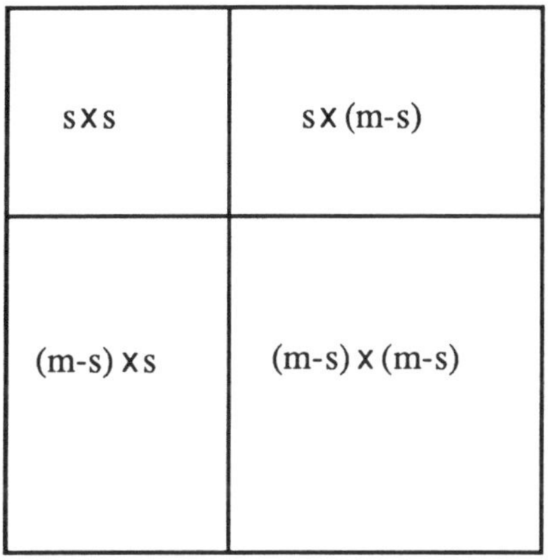

s x s	s x (m-s)
(m-s) x s	(m-s) x (m-s)

Figure 6.1. An arrangement of neurons on the mxm chessboard used in the proof of Theorem 6.3.

Equation (6.4) implies that every x_i must be an integral multiple of b_i, and so to be in transition, x_i must actually be 0. If i is the label of such a diagonal position, then (6.4) also implies exactly one neuron, neuron j, in A_i is on; all other neurons in A_i are off. Now neuron j cannot lie in the upper left sxs block of squares because then (6.4) would imply $x_j \leq 0$. So by relabelling we can assume that the s on neurons associated with the transitional neurons are all in the upper right sx(m-s) block and that all the neurons in the lower left (m-s)xs block are off. Since at least s neurons are on in the upper right block and since no two on neurons can lie in the same column, $m-s \geq s$. Furthermore, from (6.4) and the fact that each $x_j > 0$ we see that in A_j all neurons are off except for the associated neuron in transition. This forces at least s columns in the lower right $(m-s)\times(m-s)$ block to be all off. Thus at most m-s-s rows in the lower right block can have on neurons. Considering that all the neurons in the lower left block are off, there must be among the lower m-s rows at least one all off row. Select a neuron k in such a row and in the lower left block. From (6.4) we see that $x_k = .5b_k > \varepsilon$, so neuron k cannot be off. This contradiction shows that if a constant trajectory for (6.3) exists with a transitional neuron, then there must be at least two transitional neurons lying in the same row or column.   QED

THEOREM 6.4. (from the dissertation of G. Tagliarini [T]). Suppose a constant trajectory for (6.3) has two or more transitional neurons (labelled i and j) in the same row or column subset. Suppose $(2\varepsilon)^{-1}(2\varepsilon)^{-1} > b_ib_jn(n-1)/2$ for all i, j. Then the constant trajectory is unstable.

PROOF. Pairs of such neurons i and j give rise to pairs of entries L_{ij} and L_{ji} in the linear approximation matrix L of the set selection model at x (note that L is not $m\times m$ like the chessboard, but $m^2\times m^2$). In fact, $L_{ij} = -b_ig_j{}'$ and $L_{ji} = -b_jg_i{}'$. It follows that in the characteristic polynomial $\det(zI-L)$ of L the coefficient of z^2 is

$$\binom{n}{2} - \sum_{\text{all such pairs}} b_jg_i\, b_i\, g_j{}'$$

Thus the coefficient of z^2 is negative if some such pair has $(2\varepsilon)^{-1}(2\varepsilon)^{-1} > b_ib_jn(n-1)/2 = b_ib_jm^2(m^2-1)/2$. It follows that the linear approximation matrix has an eigenvalue with positive real part and x as a trajectory is unstable [W p. 66].   QED

In view of the above, for any m^2 and $\{b_i\}$ it is possible to choose ε sufficiently small so that every trajectory for (6.3) is or asymptotically approaches a constant trajectory and every stable constant trajectory is an answer set (the g matrix is an $m\times m$ permutation matrix). In particular, if all b_i = 1, then (6.3) could be used to correct errors in a *Viterbi code*, a code using the mxm permutation matrices as code words among all mxm matrices with entries 0 or 1 (see §5, chapter 4). However, for large m this code is quite sparse. Even for m = 4 this code has only 24 code words of 65536 possible binary 16-strings.

6.2 Optimization in a Modified Permutation Matrix Selection Model

We have seen that with sufficiently high gains (sufficiently small ε), answer sets (mxm permutation matrices) are the only stable attractors for (6.3). A constant trajectory x corresponding to an answer set has each $x_i = \pm b_i$ where b_i is the benefit (> 0) of having neuron i on. Let us define the *total benefit* of such an answer set to be the sum of those benefits corresponding to neurons which are on. We proceed to modify the model (6.3) so that the process of choosing a permutation matrix from randomly generated initial conditions favors anwer sets with optimal (maximal) or near optimal total benefit.

Let us initialize all neurons off but near the origin. Selecting uniformly random initial values each between -ε and -2ε suffices.

We also modify (6.3) to

$$dx_i/dt = -x_i + 1 + (2b_i-1)\exp(-t) - \sum_{j \in A_i} g_j(x_j) \qquad (6.5)$$

Here each function $1+(2b_i-1)\exp(-t)$ is initially $2b_i$ and declines exponentially to 1.

For a specific m and $\{b_i\}$ some experimentation is required to find Δt and ε for prompt convergence. As an example of such a choice, suppose the array is 10x10 and the benefits are given as follows.

1	1	1	1	1	1	1	1	1	1
2	2	2	2	2	2	2	2	2	1
3	3	3	3	3	3	3	3	1	1
4	4	4	4	4	4	4	1	2	1
5	5	5	5	5	5	1	2	1	1
6	6	6	6	6	1	2	3	2	1
7	7	7	7	1	2	3	1	1	1
8	8	8	1	2	3	4	2	2	1
9	9	1	2	3	4	1	3	1	1
10	1	2	3	4	5	2	1	2	1

It turns out that selecting $\Delta t = .1$ and $\varepsilon = .1$ (larger than that predicted in Theorem 6.3) leads to a dynamical system which converges to an answer in about 50 time steps. A spreadsheet implementation of this problem is shown on the following pages. The answer set of turned on neurons is in the A16.J25 block. Note that the calculation was terminated as soon as a permutation matrix for g values had been reached. Additional time steps would be required for x values to asymptotically approach ± 1.

Using other 10x10 benefit arrays with $\{b_i\}$ of the same magnitude, $1 \le b_i \le 10$, does not seem to affect convergence speed to an optimal or near optimal answer.

There is a pressing need somehow to make the selection of Δt and ε deterministic.

	A	B	C	D	E	F	G	H	I	J	K
1	delta t =	0.1		epsilon =		0.1	pert =	0.0101			
2	rand start =		-0.025	curr total bene =		55					
3	start cell =		-1.535	counter=	4.7						
4	x =										
5	-1.726	-1.442	-1.333	-1.249	-1.146	-1.253	-1.076	-0.988	-0.944	1.0431	
6	-1.649	-1.365	-1.256	-1.172	-1.069	-1.176	-0.999	-0.856	1.1185	-0.972	
7	-1.618	-1.333	-1.225	-1.141	-1.038	-1.144	-0.773	1.104	-1.053	-1.05	
8	-1.454	-1.168	-1.055	-0.959	-0.853	-0.953	1.1774	-1.043	-0.892	-0.996	
9	-1.513	-1.222	-1.097	-0.966	-0.66	0.9292	-1.31	-1.115	-1.178	-1.175	
10	-1.51	-1.205	-1.045	-0.584	1.0143	-1.615	-1.332	-1.129	-1.199	-1.304	
11	-1.423	-1.081	-0.526	0.9369	-1.59	-1.589	-1.296	-1.432	-1.388	-1.385	
12	-1.264	-0.469	0.8902	-1.763	-1.552	-1.548	-1.212	-1.394	-1.35	-1.455	
13	-0.328	0.7743	-1.962	-1.77	-1.556	-1.507	-1.705	-1.375	-1.573	-1.57	
14	0.917	-1.75	-1.535	-1.338	-1.053	-0.607	-1.278	-1.297	-1.144	-1.25	row
15	gain =									row sums =	
16	0	0	0	0	0	0	0	0	0	1	1
17	0	0	0	0	0	0	0	0	1	0	1
18	0	0	0	0	0	0	0	1	0	0	1
19	0	0	0	0	0	0	1	0	0	0	1
20	0	0	0	0	0	1	0	0	0	0	1
21	0	0	0	0	1	0	0	0	0	0	1
22	0	0	0	1	0	0	0	0	0	0	1
23	0	0	1	0	0	0	0	0	0	0	1
24	0	1	0	0	0	0	0	0	0	0	1
25	1	0	0	0	0	0	0	0	0	0	1
26	column sums =										
27	1	1	1	1	1	1	1	1	1	1	
28	given benefits =										
29	1	1	1	1	1	1	1	1	1	1	
30	2	2	2	2	2	2	2	2	2	1	
31	3	3	3	3	3	3	3	3	1	1	
32	4	4	4	4	4	4	4	1	2	1	
33	5	5	5	5	5	5	1	2	1	1	
34	6	6	6	6	6	1	2	3	2	1	
35	7	7	7	7	1	2	3	1	1	1	
36	8	8	8	1	2	3	4	2	2	1	
37	9	9	1	2	3	4	1	3	1	1	
38	10	1	2	3	4	5	2	1	2	1	
39	current benefits of on neurons =										
40	0	0	0	0	0	0	0	0	0	1	
41	0	0	0	0	0	0	0	0	2	0	
42	0	0	0	0	0	0	0	3	0	0	
43	0	0	0	0	0	0	4	0	0	0	
44	0	0	0	0	0	5	0	0	0	0	
45	0	0	0	0	6	0	0	0	0	0	
46	0	0	0	7	0	0	0	0	0	0	
47	0	0	8	0	0	0	0	0	0	0	
48	0	9	0	0	0	0	0	0	0	0	
49	10	0	0	0	0	0	0	0	0	0	
50											
51											
52											

	A	B	C	D	E	F	G	H	I	J	K
53	next x values =										
54	-1.652	-1.396	-1.299	-1.223	-1.13	-1.226	-1.068	-0.988	-0.949	1.0398	
55	-1.581	-1.325	-1.228	-1.152	-1.059	-1.155	-0.996	-0.867	1.1096	-0.974	
56	-1.551	-1.295	-1.198	-1.122	-1.029	-1.125	-0.791	1.0986	-1.047	-1.044	
57	-1.402	-1.145	-1.042	-0.956	-0.861	-0.951	1.1667	-1.038	-0.899	-0.995	
58	-1.453	-1.19	-1.079	-0.961	-0.685	0.9453	-1.278	-1.1	-1.159	-1.157	
59	-1.448	-1.173	-1.029	-0.615	1.0239	-1.553	-1.295	-1.111	-1.176	-1.273	
60	-1.368	-1.059	-0.561	0.9563	-1.53	-1.527	-1.261	-1.388	-1.349	-1.346	
61	-1.222	-0.507	0.9162	-1.685	-1.494	-1.488	-1.184	-1.352	-1.312	-1.408	
62	-0.378	0.8139	-1.865	-1.69	-1.495	-1.449	-1.633	-1.333	-1.514	-1.512	
63	0.9444	-1.674	-1.478	-1.299	-1.041	-0.637	-1.247	-1.266	-1.126	-1.224	
64	loop =										
65	-1.652	-1.396	-1.299	-1.223	-1.13	-1.226	-1.068	-0.988	-0.949	1.0398	
66	-1.581	-1.325	-1.228	-1.152	-1.059	-1.155	-0.996	-0.867	1.1096	-0.974	
67	-1.551	-1.295	-1.198	-1.122	-1.029	-1.125	-0.791	1.0986	-1.047	-1.044	
68	-1.402	-1.145	-1.042	-0.956	-0.861	-0.951	1.1667	-1.038	-0.899	-0.995	
69	-1.453	-1.19	-1.079	-0.961	-0.685	0.9453	-1.278	-1.1	-1.159	-1.157	
70	-1.448	-1.173	-1.029	-0.615	1.0239	-1.553	-1.295	-1.111	-1.176	-1.273	
71	-1.368	-1.059	-0.561	0.9563	-1.53	-1.527	-1.261	-1.388	-1.349	-1.346	
72	-1.222	-0.507	0.9162	-1.685	-1.494	-1.488	-1.184	-1.352	-1.312	-1.408	
73	-0.378	0.8139	-1.865	-1.69	-1.495	-1.449	-1.633	-1.333	-1.514	-1.512	
74	0.9444	-1.674	-1.478	-1.299	-1.041	-0.637	-1.247	-1.266	-1.126	-1.224	

6.3 The Quadratic Assignment Problem

It is perhaps important to mention a problem which simple neural network models seem powerless to solve well. The quadratic assignment problem can be thought of as an assignment of m machines to m possible positions. Hence a solution is an m×m permutation matrix P. Between every pair of positions is a distance, formalized as an m×m matrix D with zero diagonal entries and positive, symmetric off-diagonal entries. Between every pair of machines is a traffic flow, formalized as an m×m matrix F with zero diagonal entries and nonnegative, symmetric off-diagonal entries. The total cost C of traffic between machines is a sum of the products of all flows and distance for a chosen assignment given by

$$C = \sum_{i,j,k,l=1}^{m} F_{ij} P_{ik} P_{jl} D_{kl}$$

Considering the previous section, one might try to find good solutions by starting at a random point near the origin of m^2-dimensional space and allowing a neural network related to (6.2) or (6.3) to converge a permutation matrix. Along the way one might penalize the growth of those neurons (entries in P_{ij}) which unduly add to C. Of possible use in formulating the right side of the model are C itself and the matrix of partial derivatives of C with respect to P_{ij}. We note

$$\frac{\partial C}{\partial P_{ij}} = 2 \sum_{i,j,k=1}^{m} F_{ij} P_{jk} D_{ki}$$

However, even solving a 5x5 version of the quadratic assignment problem seems to defy simple neural network models. Suppose we are given

$$D = \begin{pmatrix} 0 & 1 & 1 & 2 & 3 \\ 1 & 0 & 2 & 1 & 2 \\ 1 & 2 & 0 & 1 & 2 \\ 2 & 1 & 1 & 0 & 1 \\ 3 & 2 & 2 & 1 & 0 \end{pmatrix}$$

and

$$F = \begin{pmatrix} 0 & 5 & 2 & 4 & 1 \\ 5 & 0 & 3 & 0 & 2 \\ 2 & 3 & 0 & 0 & 0 \\ 4 & 0 & 0 & 0 & 5 \\ 1 & 2 & 0 & 5 & 0 \end{pmatrix}$$

The optimal solution is

$$P = \begin{pmatrix} 0 & 0 & 1 & 0 & 0 \\ 0 & 0 & 0 & 1 & 0 \\ 0 & 0 & 0 & 0 & 1 \\ 1 & 0 & 0 & 0 & 0 \\ 0 & 1 & 0 & 0 & 0 \end{pmatrix}$$

with $C = 25$. Following the methods of the previous section it is not difficult to build a model which will consistently choose P with $C < 35$.

It should be noted that this test problem is particularly difficult in the sense that good solutions are generally not near (in Hamming distance) the best solution.

It would be a considerable achievement to devise a model of any sort giving near optimal solutions for the quadratic assignment problem for modest values of m, say, m = 20.

Appendix A--An Introduction to Dynamical Systems

This book requires of the reader an understanding of the basic notions of calculus and linear algebra as conventionally taught in first- and second-year mathematics courses. The following Appendix includes a few such notions which are especially relevant to neural network modeling as well as the elements of dynamical system theory.

A.1 Elements of Two-Dimensional Dynamical Systems

Suppose the state of a system is characterized at time t by two numbers, x and y, that is, $x(t)$ and $y(t)$. We also need a pair of functions $f(x,y)$ and $g(x,y)$ which are the *rates of change of the system*. Then the general mathematical formulation for a *two-dimensional difference equation dynamical system* is

$$x(t+\Delta t) = x(t) + f(x,y)\Delta t \qquad\qquad (A.1)$$
$$y(t+\Delta t) = y(t) + g(x,y)\Delta t$$

The first equation means: to find the next x value, the value of x at time $t+\Delta t$, add to the old x value, namely $x(t)$, the product of the local rate of change $f(x,y)$ and the time step (one nanosecond, one day, or whatever other time interval is suitable for the model at hand). Similarly, the new value of y at time $t+\Delta t$ is obtained by adding to the current value $y(t)$ the product of the local rate of change per time step $g(x,y)$ and the time step Δt itself. We call *state space* the set of all states which are valid in the system. The set of all valid pairs of points from state space and time is called *state-time*.

Suppose state space for a particular model is defined to be the set of nine points (x,y) where $x,y \in \{1,2,3\}$. A special case of (A.1) can be given by a list of system function values written in function form as follows:

f(x,y) = +1 at (1,1)   (A.2)

      = -1 at (2,3) and (3,3)

      = 0 at all other points

g(x,y) = +1 at (2,1), (3,1), and (3,2)

      = -1 at (1,2) and (1,3)

      = 0 at all other points

We can now compute a trajectory for (A.2) with $\Delta t = 1$ starting at the initial state (3,1). The trajectory consists of the sequence of points

(3,1)→(3,2)→(3,3)→(2,3)→(1,3)→(1,2)→(1,1)→(2,1)→(2,2)→(2,2)→...

This trajectory is shown in Fig. A.1.

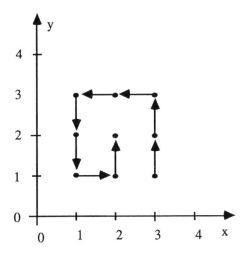

Figure A.1. A trajectory for a discrete two-dimensional difference equation dynamical system (A.2).

By contrast, the trajectory for (A.2) starting at (2,2) is

(2,2)→(2,2)→(2,2)→(2,2)→(2,2)→(2,2)→(2,2)→(2,2)→(2,2)→(2,2)→...

This trajectory is called a *constant trajectory* because it never changes with time, that is, the rates of change f(2,2) and g(2,2) are both 0. Moreover, this particular constant trajectory is called an *attractor trajectory* because after a certain amount of time (determined by initial conditions) any other trajectory starting in state space will be "arbitrarily close" to (2,2) ("within epsilon for an arbitrarily small, positive epsilon," as mathematicians enjoy saying, using the standard measure of distance in 2-dimensional space).

Suppose we have instead of a clue list the mathematical rules: $\Delta t = .001$; $f(x,y) = 200x - x^2 - 10xy$; and $g(x,y) = xy - 10y^2$; or the equivalent as equations:

$$x(t+.001) = x(t)+[200x(t)-x^2(t)-10x(t)y(t)](.001) \tag{A.3}$$
$$y(t+.001) = y(t)+[x(t)y(t)-10y^2(t)](.001)$$

The reader should be able to show that (100,10) is a constant trajectory for system (A.3).

Next consider the trajectory for (A.3) starting at $t = 0$, $x(0) = 80$, $y(0) = 15$. The trajectory values for the first few time steps are

$(x(0), y(0)) = (80,15)$
$(x(.050), y(.050)) \cong (100.1128, 10.0375)$
$(x(.100), y(.100)) \cong (100.0024, 10.0001)$
$(x(.150), y(.150)) \cong (100.0000, 10.0000)$

A sketch of points of the trajectory is shown in Fig. A.2. The points are connected for the sake of visual clarity.

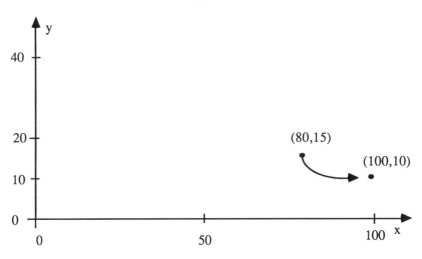

Figure A.2. A trajectory for the two-dimensional difference equation dynamical system (A.3).

This model arises in ecology and is a simplistic version of prey (x) and predator (y) dynamics.

Figure A.2 obviously suggests a sort of *stability*: the constant trajectory (100,10) is the treasure of the dynamical system treasure hunt which starts at (80,15). There is a perverse element to the mathematics, however, in that the treasure is never actually reached. Sooner or later the modeler must say that the trajectory for all practical purposes has reached the treasure. The modeler, moreover, might compute other trajectories starting near (100,10), say, starting at (115,5) or (120,20). Each such trajectory will likewise seem to approach asymptotically the constant trajectory (100,10), so the modeler is tempted to assert that the equations are stable in some sense.

Fortunately there is a mathematical line of reasoning which shows that all trajectories for (A.3) starting in the positive orthant must asymptotically approach the constant trajectory (100,10). That theory is the stability theory of M. A. Lyapunov, to be outlined below.

A.2 Elements of n-Dimensional Dynamical Systems

The reader can no doubt appreciate that dynamical systems might use many more than two state variables. In a model of large dimension there would not be enough letters in the English or any other Western alphabet to assign one letter to each compartment as its algebraic label. So we resort to subscripts, numbers attached to a general component symbol x. We write $x = (x_1,x_2,x_3,...,x_n)$ and so x becomes a shorthand for the whole n-vector. This leads to the general difference equation for x_i :

$$x_1(t+\Delta t) = x_1(t)+f_1(x_1(t),x_2(t),...,x_n(t))\Delta t \qquad (A.4)$$
$$x_2(t+\Delta t) = x_2(t)+f_2(x_1(t),x_2(t),...,x_n(t))\Delta t$$

...

$$x_n(t+\Delta t) = x_n(t)+f_n(x_1(t),x_2(t),...,x_n(t))\Delta t$$

The system (A.4) is just the n-dimensional extension of the two-dimensional system (A.1). Generally speaking, computing trajectories for some version of (A.4) would be conceptually simple but arithmetically tedious, just the thing for a computing machine.

Much of mathematics amounts to the creation and manipulation of mathematical shorthand. Therefore we might as well also abbreviate in (A.4) the *rate of change functions* $(f_1,f_2,f_3,...,f_n)$ = f so the whole *difference equation dynamical system* can be written as

$$x(t+\Delta t) = x(t)+f(x(t))\Delta t \qquad (A.5)$$

Some neural network models might be time driven by external factors. In such cases it would be desirable to incorporate time dependent factors explicitly and directly in the rate of change function f. Generalizing one last time, we write the *time dependent difference equations*

$$x(t+\Delta t) = x(t)+f(x(t),t)\Delta t \qquad (A.6)$$

A.3 The Relation Between Difference and Differential Equations

As everyone who has studied calculus knows, the *derivative of a function* is approximated by the slope of the hypotenuse of a little triangle, in fact, an arbitrarily small triangle. It seems that the concept that either makes or breaks the calculus student is this: even though the sides of the triangle can become arbitrarily small, the ratio of the length of the vertical side to the length of the horizontal side generally becomes, for a differentiable function, not zero but a particular number, the *rate of change* or *derivative* of the function. The triangle we have in mind is shown in Fig. A.3.

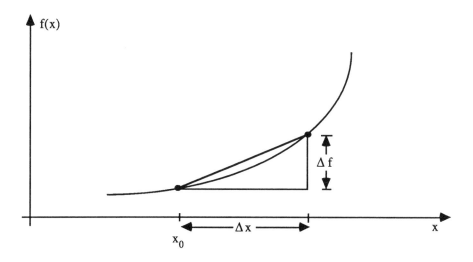

Figure A.3. The great shrinking triangle of differential calculus.

One must imagine a triangle shrinking so that the ratio of lengths of its sides approaches a fixed number, even though the lengths themselves approach zero. Generally in computing derivatives of (differentiable) functions, the triangle changes shape slightly and ever more slowly as it shrinks.

The symbolisms

$$df/dx = \lim_{\Delta x \to 0} \frac{\Delta f}{\Delta x}$$

and

$$\dot{x}(t) = dx/dt = \lim_{\Delta t \to 0} \frac{\Delta x}{\Delta t}$$

are undoubtedly familiar. Note that the dot notation always refers to differentiation with respect to time t.

The difference $x(t+\Delta t)-x(t)$ is the change in x value from t to $t+\Delta t$. Shorthand for this difference is $\Delta x(t)$. Thus associated with the difference equation dynamical system $x_i(t+\Delta t) = x_i(t)+f_i(x,t)\Delta t$ are the *rate of change functions* $f_i(x,t)$.

The *differential equation dynamical system* analog of (A.6) is

$$dx_i/dt = f_i(x,t) \tag{A.7}$$

or, as n-vectors,

$$dx/dt = f(x,t) \tag{A.8}$$

In differential equation dynamical systems, the functions in $f(x,t)$ in (A.7) and (A.8) are called the *system functions*. An n-vector of functions $x(t)$ satisfying the equations is called a *trajectory of the dynamical system.*

Of course, if we start with a difference equation dynamical system, there is always a way--namely, brute force--to determine trajectories of the system. Differential equation dynamical systems are more subtle. Let us refer to a central theorem on the existence of trajectories of differential equation dynamical systems.

THEOREM A.1--Trajectory Existence Theorem. Suppose the system functions $f(x,t)$ for the dynamical system $dx/dt = f(x,t)$ are continuous and have continuous derivatives in some open set of state-time. Then to any initial state in that open set is associated a local unique trajectory $x(t)$ solving $dx/dt = f(x,t)$.

The Trajectory Existence Theorem is a standard result in dynamical system theory and can be studied in, for example, [S, p. 8].

With the increasing availability of computers and their use by nonmathematicians, it seems to be ever easier to be successful at just about anything except pure mathematics while believing in the existence of only a finite number of numbers. If one uses only the decimal numbers with, say, 99 digits to the left and 99 digits to the right of the decimal, together with the additive inverses (negatives) of such numbers, one can do much science and engineering with a computer . It is logical to question, then, the necessity of studying the *differential equation* as opposed to the conceptually simpler *difference equation* viewpoint.

Studying differential equations is necessary in modeling because the generic essence of a model is often revealed only in the limit $\Delta t \to 0$, not with numerous experimental choices of "small" Δt. Qualitative knowledge based on mathematical ideas of how systems work is without a doubt a modeling tool just as useful as computer simulations.

Let us consider an example of such usefulness. Consider the two-dimensional system

$$dx_1/dt = x_2 \qquad\qquad\qquad\qquad (A.9)$$
$$dx_2/dt = -x_1$$

The two system functions are

$$f_1(x_1,x_2,t) = x_2$$
$$f_2(x_1,x_2,t) = -x_1$$

These system functions obviously fulfill the conditions in the Trajectory Existence Theorem. The reader should verify by differentiation that

$$x_1(t) = A \cos(t) + B \sin(t) \qquad\qquad\qquad (A.10)$$
$$x_2(t) = -A \sin(t) + B \cos(t)$$

is a trajectory for the system (A.9) where A and B are arbitrary constants. Clearly A and B are determined by initial conditions. If the initial time is $t = 0$, then $x_1(0) = A$ and $x_2(0) = B$. If $A = B = 0$, then we have a constant trajectory at (0,0). For other values of A and B we note that generally $x_1(t)^2 + x_2(t)^2 = A^2+B^2$, a constant. Thus the reader should be able to verify that any trajectory for (A.10) traces out a circle of radius $(A^2+B^2)^{.5}$ centered at (0,0) and passes through its initial point every 2π time units. A typical trajectory is shown in Fig. A.4.

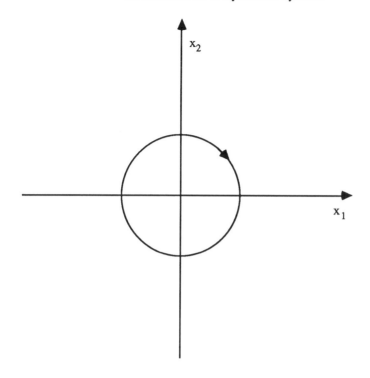

Figure A.4. A typical trajectory for the system (A.9).

Trajectories of the difference equation analogue of the system would not, of course, be points exactly on a circle and would generally never pass through a given point more than once. In fact, points of such trajectories would lie on curves which would spiral away from (0,0), regardless of how small the choice of Δt. Over the long run, there is a big difference between a trajectory which goes around and around a circle and a trajectory which spirals outward! This example warns us of potential modeling subtleties.

Fortunately the concept of *stability* simplifies the transition between differential and difference equations. Roughly speaking, if a model has this property, stability, then whenever Δt is sufficiently small or smaller, computer simulations do give an accurate and consistent qualitative picture of system dynamics.

The method of generating trajectories in (A.7) is called the *forward-difference procedure* or *Euler method*. There are many more sophisticated numerical schemes for generating trajectories and such methods are indispensable in many types of mathematical modeling. However, experience

to date has shown that in neural network models nothing is gained by using
such methods.

A.4 The Concept of Stability

Suppose a train proceeds down a track toward its destination. Suppose
the remaining distance to its destination is always decreasing by at least 100
km/hr so long as the train is at least 10 km from its destination. Suppose the
initial distance from the train to the destination is 510 km. The reader can no
doubt appreciate that the train can take at most five hours before lying within
10 km of its destination. This simple idea could readily be extended to other
dimensions, say the three-dimensional rendezvous of a spacecraft with a
satellite. The details of motion are unknown and not needed, yet the
conclusion is obvious.

Next suppose the distance to the destination is not always decreasing,
or at least not known to be always decreasing in a simple way. Suppose
instead that "distance to destination" is decreasing according to some observer
viewing the progress of the vehicle through some sort of imperfect lens. The
imperfect lens makes straight lines wavy lines, but somehow one can be sure
that, say, 100 km reported by the observer is in reality at least 90 km and not
more than 110 km. It seems reasonable to assume that upper limits on the time
required to enter a neighborhood of the destination could still be calculated.

Such is the essential idea of Lyapunov's theory of stability of motion:
details of motion are not essential to guarantee the approach of a destination
so long as some measure of the remaining distance is decreasing. Lyapunov's
results make this idea mathematically precise.

The first step in supplying that precision is the notion of distance in
n-space. The *distance d(x,y) between two points in n-space* is defined as

$$d(x,y) = \left[\sum_{i=1}^{n} (x_i - y_i)^2 \right]^{.5} \tag{A.11}$$

Thus for example the distance between (1, -1, 2, 0, 3) and (0, -2, 2, 0, 1) in
five-dimensional space is $(6)^{.5}$.

An *attractor trajectory* $\underline{x}(t)$ is a trajectory for a dynamical system with
two properties:

First there must be $\alpha > 0$ such that any other trajectory $x(t)$ with $d(x(t_0),\underline{x}(t_0)) < \alpha$ at any initial time t_0 must asymptotically approach $\underline{x}(t)$ as $t \to \infty$.

Second, for any $\varepsilon > 0$ there must exist $\delta > 0$, δ determined by ε only, such that $d(x(t_0),\underline{x}(t_0)) < \delta$ implies $d(x(t),\underline{x}(t)) < \varepsilon$ for all $t > t_0$.

Put another way, any trajectory which starts sufficiently close to $\underline{x}(t)$ must asymptotically approach $\underline{x}(t)$; and trajectories can be guaranteed to stay close to $\underline{x}(t)$ (no wild transient behavior before approaching $\underline{x}(t)$)by just requiring that they start close to $\underline{x}(t)$.

A set in n-space has *finite diameter* if there is a finite number D such that $d(x,y) \le D$ for any pair of points x and y in the set. An *attractor region* **A** for a dynamical system is a subset of state space which has finite diameter and one property relative to the system. With respect to the system, **A** must satisfy the following condition: there must exist $\delta > 0$ such that any trajectory $x(t)$ starting at time t_0 within a distance δ of **A** must must subsequently lie in **A** for all times $t > T+t_0$ where T is determined by δ and t_0 only. (Starting at time t_0 at $x(t_0)$ within a distance δ of **A** means $d(x(t_0),\underline{x}) < \delta$ for some $\underline{x} \in$ **A**.)

In this book we assume that the concept *stability of a dynamical system* has two parts. First it means that there is an attractor trajectory or attractor region for the system. This leads to the second notion, the size of the *region of attraction* of the attractor trajectory or region. By region of attraction we mean the set of points in state-time (thought of as the initial values of trajectories) which asymptotically approach the attractor trajectory or which enter the attractor region after a finite time interval. Certainly more than one attractor trajectory or attractor region might exist; in that case, a description of how state-time is partitioned into regions of attraction is needed.

As an example of the partitioning of state space into regions of attraction, let us consider the following trajectory diagram from Chapter 1.

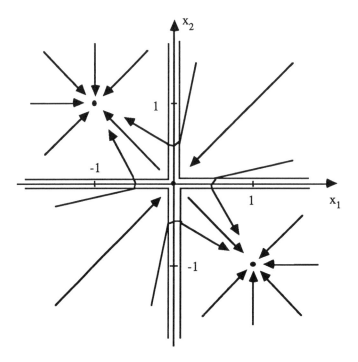

Figure A.5. Typical trajectories for a memory model in two-dimensional space with two memories.

In Fig. A.5 the region in (x_1, x_2)-space with x_2-$x_1 > 0$ is the region of attraction of $(-1,1)$; the region with x_2-$x_1 < 0$ is the region of attraction of $(1,-1)$; and the region with $x_2 = x_1$ is the region of attraction of $(0,0)$.

A.5 Limit cycles

An important feature of attractor trajectories generally is the possibility of time dependence. A nonconstant attractor trajectory serves as a sort of moving target pursued by the model, as opposed to a constant attractor trajectory, a stationary target.

In some models, including some without time dependence in their system functions, another type of moving target can occur. A *limit cycle* is a simple closed loop in state space with two properties. First, any trajectory starting at any time at any point on the limit cycle stays on the limit cycle forever and is a cyclic trajectory with period T, T being a characteristic of the limit cycle. Second, any trajectory starting sufficiently close to the limit cycle must asymptotically approach some trajectory in the limit cycle.

The following abstract dynamical system has a limit cycle:

$$dx_1/dt = (1-x_1^2-x_2^2)x_1 + x_2 \qquad\qquad (A.12)$$
$$dx_2/dt = (1-x_1^2-x_2^2)x_2 - x_1$$

It turns out that the unit circle is a limit cycle for this system. Also, $x(t) = (0,0)$ is a constant trajectory. Typical trajectories for (A.12) are shown in Fig. A.6.

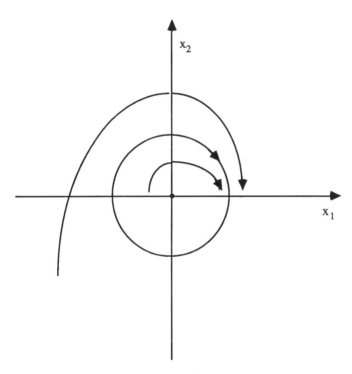

Figure A.6. Trajectories of a two-dimensional dynamical sytem (A.12) with a limit cycle.

The key feature in Fig. A.6 is that trajectories other than (0,0) and those which start on the unit circle asymptotically approach some cyclic trajectory which lies in the unit circle. Different initial values generally result in the asymptotic approach of different cyclic trajectories (moving targets). Note that distinct trajectories starting on the unit circle remain forever separated by a fixed distance; thus there is no attractor trajectory for the system.

A.6 Lyapunov Theory

Euclidean distance in (A.11) is not always the "best" way to measure how close a state x is to the current value of an attractor trajectory $\underline{x}(t)$. Instead, "closeness" can be measured by a Lyapunov function $\Lambda(x,t)$, as we now illustrate.

A trajectory starting at (2,1) for the dynamical system

$$dx_1/dt = x_1(-.1x_1-2x_2+2.1)$$
$$dx_2/dt = x_2(2x_1-.1x_2-1.9)$$

$\hspace{9cm}$ (A.13)

is shown in Fig. A.7.

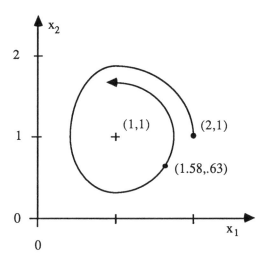

Figure A.7. A trajectory of the two-dimensional model (A.13).

Although the trajectory in Fig. A.7 seems to approach asymptotically the constant trajectory (1,1), we note that at certain points the distance between the trajectory and (1,1) is apparently increasing. In fact, the trajectory starting at (2,1) passes close to (1.58,.63) whereupon the rate of change of distance is about +.50 .

However, let us consider another way to measure the current state of the system relative to $\underline{x}(t) = (1,1)$. Let the function $\Lambda(x_1,x_2)$ be defined by

$$\Lambda(x_1,x_2) = x_1 - \ln(x_1) - 1 + x_2 - \ln(x_2) - 1 \qquad (A.14)$$

(In the Cyrillic alphabet, Λ is the first letter of "Lyapunov" and is pronounced "ell.")

Along any trajectory $x(t)$ of (A.13) in the positive orthant, the rate of change of Λ (total derivative of Λ with respect to time) is

$$\begin{aligned} d\Lambda/dt &= dx_1/dt - x_1^{-1}dx_1/dt + dx_2/dt - x_2^{-1}dx_2/dt \qquad (A.15)\\ &= (x_1-1)x_1^{-1}dx_1/dt + (x_2-1)x_2^{-1}dx_2/dt\\ &= -.1(x_1-1)^2 - .1(x_2-1)^2 \end{aligned}$$

Thus along the constant trajectory $\underline{x}(t) = (1,1)$, Λ is a constant, zero. <u>Along all other trajectories, (A.15) implies that Λ always decreases.</u>

To any nonnegative number λ, there corresponds a *level set,* the set of all points in space for which the value of the function Λ is λ.

The relationship between the system (A.13) and the level sets of above Λ is shown in Fig. A.8.

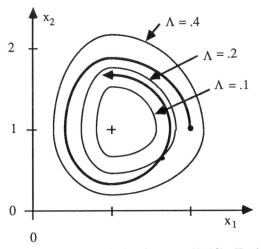

Figure A.8. An analysis of system (A.13). Typical relations of level sets of the Lyapunov function (A.14) are shown, namely, the level sets with $\Lambda = .1, .2,$ and $.4$. The constant attractor trajectory at (1,1) is shown as the mark +. A typical trajectory which asymptotically approaches the constant trajectory is shown as a bold line with a point.

The above level sets suggest the concepts of "inside" and "outside" relative to a trajectory $\underline{x}(t)$ such as the constant trajectory at $(1,1)$. Suppose there are two level sets L_1 and L_2 in state-time with the property that L_1, L_2, and $\underline{x}(t)$ have no points in common. We say L_1 is *inside* L_2 (and L_2 is *outside* L_1) if each ray beginning at every point $\underline{x}(t)$ and perpendicular to the time axis first intersects L_1 and then L_2. Thus in Fig. A.8, the level set L_1 where $\Lambda = .1$ is inside the level set L_2 where $\Lambda = .2$.

If L_1, L_2, ..., L_q are all level sets of some function about $\underline{x}(t)$ with L_i inside L_{i+1} for each $i = 1, 2, ..., q-1$, then we say the level sets are *nested* about $\underline{x}(t)$.

Let R denote a region of state-time. We shall assume that R is at least big enough to include all level sets $\Lambda = \lambda$ of Λ for $0 \le \lambda \le \delta$ for some positive number δ. We shall also use the level set of radius ε (*level cylinders*) about $\underline{x}(t)$, meaning for each time t the set of all points in state space at a distance ε from $\underline{x}(t)$.

We are now in a position to prove a version of M. A. Lyapunov's important result.

THEOREM A.2--Lyapunov Theorem. Suppose we have a dynamical system $dx/dt = f(x,t)$ in a region R of n-dimensional space with trajectory $\underline{x}(t)$ and an analytic nonnegative function $\Lambda(x,t)$, all related as follows:

    1. for any level set L contained in R, there are level cylinders inside and outside L.

    2. for any level cylinder C, there is a level set L_1 of Λ inside C; if the radius of C is smaller than a certain number determined by $\underline{x}(t)$, then C is inside some level set of Λ in R;

    3. given a level cylinder in R, there is a positive number β such that $d\Lambda/dt \le -\beta$ at all points in R and outside the level cylinder.

If these three conditions are met, then $\underline{x}(t)$ is an attractor trajectory for the dynamical system.

PROOF. First we show that $\underline{x}(t)$ is asymptotically approached by all nearby trajectories. Let ε be any positive number and let $x(t_0)$ in R be the initial state of an arbitrary trajectory $x(t)$. Suppose $x(t_0)$ is not already inside the level

cylinder C of radius ε. Our task is to show that x(t) will be inside C after a
finite time interval determined by the distance between $x(t_0)$ and $\underline{x}(t_0)$.

   Let L denote the level set of Λ which contains $x(t_0)$. If C is not inside L,
choose a smaller value for ε so it is. Using conditions 1 and 2 we can choose a
level set L_1 of Λ and level cylinders C_1 and C_0 such that C_1, L_1, C, L, C_0 are
nested level sets (in the given order) about $\underline{x}(t)$. This arrangement of level
sets is shown in Fig. A.9.

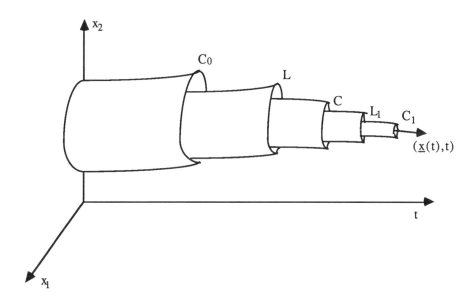

Figure A.9. Level sets and level cylinders as needed in the proof of the
Lyapunov Theorem. Only two of n spatial dimensions can be shown.

By condition 3, $d\Lambda/dt$ is less than or equal to some $-\beta$ outside C_1. Let λ be the
value of Λ on L and let λ_1 be the value of Λ on L_1. It follows that the arbitrary
trajectory can take at most $(\lambda-\lambda_1)/\beta$ time units to enter L_1. Since λ can be
bounded in terms of the radius of C_0 and since that radius is greater than the
distance from $x(t_0)$ to $\underline{x}(t_0)$, the time interval taken before the trajectory must
enter C can be bounded by the initial distance from $x(t_0)$ to $\underline{x}(t_0)$.

   Let C be a cylinder of an arbitrarily small radius about $\underline{x}(t)$. Let L_1 be a
level set of Λ and let C_1 be a level cylinder so that C_1, L_1, and C are nested
about $\underline{x}(t)$. No trajectory starting in C_1 can ever escape L_1 because $d\Lambda/dt$ is

negative on L_1; all trajectories passing through L must puncture L inwardly. Thus any trajectory starting closer to $\underline{x}(t)$ than the radius of C_1 must thereafter remain within the radius of C of $\underline{x}(t)$.

The above observations imply that $\underline{x}(t)$ is an attractor trajectory.   QED

In the case of the above two-dimensional model (A.13),

$$d\Lambda/dt = -.1(x_1-1)^2 -.1(x_2-1)^2$$

Thus outside the level cylinder of radius ε,

$$d\Lambda/dt = -.1\varepsilon^2$$

To satisfy condition 3 we need only set $\beta = .1\varepsilon^2$. Note also that R can be taken to be the entire positive orthant and all time.

A.7 The Linearization Theorem

A linear dynamical system is a dynamical system with system functions which are linear in x. For example, the following is a simple *linear dynamical system*

$$dx_1/dt = -.1x_1 - 2x_2$$
$$(A.15) dx_2/dt = 2x_1 - .1x_2$$

In fact, as we will soon see, (A.15) is the linearization of the above predator-prey dynamical system (A.13) about the constant trajectory (1,1).

Generally the *linearization* of a dynamical system

$$dx_i/dt = f_i(x,t)$$

about a trajectory $\underline{x}(t)$ is defined as the linear dynamical system

$$dx_i /dt \;\; = \;\; \sum_{j=1}^{n} \frac{\partial f_i}{\partial x_j} (\underline{x}(t),t) x_j \tag{A.16}$$

with linear approximation matrix

$$A_{ij}(t) = \frac{\partial f_i}{\partial x_j}(x(t), t)$$

If each partial derivative on $x(t)$ is actually a constant, as in our example, the linearization is called *autonomous*.

The Linearization Theorem stated below is included in this book because for some nonlinear systems it enables us to tell whether or not a given trajectory $x(t)$ is an attractor trajectory.

THEOREM A.3--Linearization Theorem. Suppose system functions $f(x,t)$ are analytic in x and t. Suppose also:

1. 0 is an attractor trajectory for the linearization (A.16) of $dx/dt = f(x,t)$ about $x(t)$; and
2. there is a number N such that the absolute value of each component of the the linear approximation matrix $|\partial f_i/\partial x_j|$ at $x(t)$ is bounded by N for all i, j, t.

Then $x(t)$ is an attractor trajectory for the full system, and some open neighborhood of $x(t)$ in state-time is a region of attraction for $x(t)$.

A proof and details of the implications of this theorem can be found in [W, Chapter 5]. An important limitation of the Linearization Theorem is that nothing is said about the size of the region of attraction of $x(t)$.

A.8 The Stability of Linear Systems

In this section we shall study a test for the stability of the linear dynamical system with constant coefficients

$$dx_i/dt = \sum_{j=1}^{n} A_{ij} x_j \qquad (A.17)$$

First we need to recall some linear algebra. An *n×n matrix A* consists of n^2 numbers (for our purposes, always real numbers) arranged in n rows and n columns. A *permutation* of the n integers 1,2,3,...,n is simply a reordering of the integers, for example, n,n-1,...,2,1. Any permutation can be rendered into a natural order by a sequence (not unique) of steps, each step being the interchange of two adjacent integers. If the number of steps in such a sequence is even (including zero), the *sign of the permutation* is +1. Otherwise the sign of the permutation is -1. Even though the sequence of steps needed to convert a given permutation into a natural order is not uniquely determined, the sign of the permutation is.

We also recall that the *determinant* of an n×n matrix A is defined as

$$\det A = \sum_{i \,\in\, \text{permutations of } 1,2,\ldots,n} \text{sign}\,(i)\, A_{1i_1} A_{2i_2} \cdots A_{ni_n}$$

where the sum is carried out over all possible permutations of 1,2,3,...,n. Thus

$$n = 1 \Rightarrow \det A = A_{11}$$
$$n = 2 \Rightarrow \det A = A_{11}\,A_{22} - A_{12}\,A_{21}$$
$$n = 3 \Rightarrow \det A =$$
$$A_{11}A_{22}A_{33} + A_{12}A_{23}A_{31} + A_{13}A_{21}A_{32} - A_{11}A_{23}A_{32} - A_{12}A_{21}A_{33} - A_{13}A_{22}A_{31}$$

Clearly the number of summands in det A is *n factorial*, $n! = n\cdot(n-1)\cdot(n-2)\ldots 3\cdot 2\cdot 1$. For larger values of n, say n = 100, calculating det A by brute force is not feasible. Of course, there are tricks which can be used for certain types of matrices; type is determined by how many zeroes and where they occur.

A *polynomial* in the single variable z is a finite sum of multiples of nonnegative integer powers of z.

The n×n *identity matrix* I has +1 in each I_{ii} *diagonal* entry and 0 in each I_{ij}, $i \neq j$ *off-diagonal* entry. Thus the 3×3 identity matrix is

$$\begin{pmatrix} 1 & 0 & 0 \\ 0 & 1 & 0 \\ 0 & 0 & 1 \end{pmatrix}$$

By zI is meant the matrix with the variable z in each diagonal entry and 0 in each off-diagonal entry.

The *characteristic polynomial* of an n×n matrix A is the determinant of the n×n matrix zI-A where $(zI-A)_{ii} = z-A_{ii}$ and $(zI-A)_{ij} = -A_{ij}$, i≠j. Thus the characteristic polynomial of

$$\begin{pmatrix} 1 & 0 & 2 \\ 2 & 0 & 0 \\ 0 & 0 & 3 \end{pmatrix}$$

is $(z-1)(z-0)(z-3) = z^3 - 4z^2 + 3z$.

The *Fundamental Theorem of Algebra* as applied in this section tells us that any polynomial

$$p(z) = z^n + p_{n-1}z^{n-1} + ... + p_2z^2 + p_1z + p_0$$

where $p_0, p_1, p_2, ...$ are real numbers always has n *roots* $r_1, r_2, ..., r_n$; each root is a number (possibly complex) satisfying $p(r_i) = 0$. Put another way,

$$p(z) = (z-r_1)(z-r_2)...(z-r_n).$$

The *eigenvalues* of an n×n matrix A are the roots of the characteristic polynomial of A.

A fundamental result of the theory of linear dynamical systems is that the origin 0 is an attractor trajectory for (A.17) if and only if each eigenvalue of A has a <u>negative</u> real part, that is, either is a negative real number or is a complex number with negative real part [W, p. 49]. If 0 is an attractor trajectory, then the region of attraction for 0 is all of n-dimensional space.

If a linear approximation matrix in (A.16) has even one eigenvalue which has a <u>positive</u> real part, then neither 0 for the linear approximation system

$$dx_i/dt = \sum_{j=1}^{n} \frac{\partial f_i}{\partial x_j} (\underline{x}(t),t)x_j \qquad (A.16)$$

nor $\underline{x}(t)$ for the full nonlinear system is an attractor trajectory [W, p.129]. Thus a purely algebraic test can sometimes determine aspects of the qualitative performance of a nonlinear dynamical system.

The *Hurwitz test* [W, p. 67-71] does not involve calculation of the actual eigenvalues of A. For low values of n the Hurwitz Test can be rewritten as follows. All the eigenvalues of an n×n matrix have negative real part if and only if

$n = 1$:	$p_0 > 0$
$n = 2$:	$p_0, p_1 > 0$
$n = 3$:	$p_0, p_1, p_2 > 0$; $p_2 p_1 - p_0 > 0$
$n = 4$:	$p_0, p_1, p_2, p_3 > 0$; $p_3 p_2 p_1 - p_1^2 - p_0 p_3^2 > 0$

For higher values of n, the Hurwitz test is not so easily expressed. The Hurwitz test is useful in determining the qualitative roles played by various coefficients in low-dimensional nonlinear systems.

Suppose an n-dimensional linear dynamical system has p eigenvalues with negative real parts (and associated p-dimensional subspace) and n-p eigenvalues with positive real parts (and associated (n-p)-dimensional subspace). Then a trajectory which starts in the p-dimensional subspace will asymptotically approach the origin. A trajectory which starts in the (n-p)-dimensional subspace will exponentially diverge from the origin. If one takes a point in the p-dimensional subspace and perturbs it by a small random amount, the subsequent trajectory will generally travel toward the origin, move slowly near the origin, then swerve away from the origin, asymptotically approaching a diverging trajectory.

Suppose all eigenvalues of a linear system have negative real parts. If we seek to destabilize the origin by changing the system structure to make exactly one eigenvalue have positive real part, all other eigenvalues unchanged, then the quantitative performance of the system will depend on the relative size of that eigenvalue. The larger the real part, the faster on average trajectories will diverge from the origin.

Appendix B--Simulation of Dynamical Systems with Spreadsheets

A spread sheet is an electronic ledger or matrix in which each entry (cell) can be filled with a written comment, a number, or a function which calculates a number in terms of numbers in the other cells.

Brute force simulation of simple dynamical systems can be accomplished by entering labels in row 1, constants and initial conditions in row 2, and one iteration in row 3. Then copying row 3 into rows 4 through 102 to simulates 100 time steps of the system.

For example, consider the damped harmonic oscillator system

$$dx_1/dt = x_2$$
$$dx_2/dt = -x_1 - x_2$$

The difference equation analog of this system is

$$x_1(t+\Delta t) = x_1(t) + (x_2(t))\Delta t$$
$$x_2(t+\Delta t) = x_2(t) + (-x_1(t)-x_2(t))\Delta t$$

Suppose we wish to calculate the trajectory of this system starting at (1,0) with $\Delta t = .1$. In a spreadsheet, we might use cells A1, B1, and C1 as labels for x_1, x_2, and Δt. Cells A2, B2, and C2 would then contain the initial conditions and the value of Δt. Cell A3 contains the formula

=A2+(B2)*$C$2

which is the spreadsheet version of $x_1(t+\Delta t) = x_1(t) + (x_2(t))\Delta t$. Cell B3 contains the formula

=B2+(-A2-B2)*$C$2

which is the spreadsheet version of $x_2(t+\Delta t) = x_2(t) + (-x_1(t)-x_2(t))\Delta t$. The entries in cells A3 and B3 are then copied down the spreadsheet to iterate the equation. Thus entries in cells A4 and B4 are computed in terms of entries in cells A3 and B3, and so on. It is the relative position in the spreadsheet which determines the next x_1 and x_2 values in terms of the previous values. Note, however, that the use of $ in $C$2 makes that portion of formula entries refer absolutely to the number in cell C2, not to a relative cell.

A spreadsheet which accomplishes this is shown on the following page.

	A	B	C
1	x1 =	x2 =	delta t =
2	1	0	0.1
3	1	-0.1	
4	0.99	-0.19	
5	0.971	-0.27	
6	0.944	-0.3401	
7	0.90999	-0.40049	
8	0.869941	-0.45144	
9	0.824797	-0.4932901	
10	0.775468	-0.5264408	
11	0.7228239	-0.5513435	
12	0.6676896	-0.5684916	
13	0.6108404	-0.5784114	
14	0.5529993	-0.5816543	
15	0.4948338	-0.5787888	
16	0.436955	-0.5703933	
17	0.3799156	-0.5570494	
18	0.3242107	-0.5393361	
19	0.2702771	-0.5178235	
20	0.2184947	-0.4930689	
21	0.1691879	-0.4656115	
22	0.1226267	-0.4359691	
23	0.0790298	-0.4046349	
24	0.0385663	-0.3720744	
25	0.0013589	-0.3387236	
26	-0.0325135	-0.3049871	
27	-0.0630122	-0.271237	
28	-0.0901359	-0.2378121	
29	-0.1139171	-0.2050173	
30	-0.1344188	-0.1731239	
31	-0.1517312	-0.1423696	
32	-0.1659682	-0.1129595	
33	-0.1772641	-0.0850667	
34	-0.1857708	-0.0588337	
35	-0.1916542	-0.0343732	
36	-0.1950915	-0.0117705	
37	-0.1962685	0.0089157	
38	-0.195377	0.027651	
39	-0.1926119	0.0444236	
40	-0.1881695	0.0592424	
41	-0.1822453	0.0721351	
42	-0.1750318	0.0831461	
43	-0.1667171	0.0923347	
44	-0.1574837	0.099773	
45	-0.1475064	0.105544	
46	-0.136952	0.1097403	
47	-0.1259779	0.1124614	
48	-0.1147318	0.1138131	
49	-0.1033505	0.113905	
50	-0.09196	0.1128495	
51	-0.080675	0.1107606	
52	-0.069599	0.107752	

Lists of numbers do not make good reading. Consequently, for models of low dimension, one might use the graphing capabilities of the spreadsheet to generate curves of x_1 versus time, x_2 versus time, x_1 versus x_2, level sets, and so on.

For neural network models with many more than two neurons, graphs are not useful. Instead the system can be studied by commanding the spreadsheet to update values in fixed cells as the system evolves. To do this, we specify labels, constants, and initial conditions as before. It is a good idea to use plenty of labels or leave empty cells as margins for active cells. The initial system values (and subsequent values) might appear in a rectangular matrix of cells called the *variable block*. For the memory model of Chapter 3 one would then compute a rectangular matrix of gain function values, making heavy use of the copy capability of the spreadsheet. Other arrays of values would be specified, finally arriving at a *new block* of system variables constituting one complete iteration.

Suppose the initial values of the system are in the rectangular array of cells with A3 and C5 at opposite corners, designated A3:C5 in Excel™. Suppose after one iteration the new values are in A43:C45 . For the purpose of iteration we need a *loop block*, here obtained by entering =A43 in A47 and copying throughout A47:C49. Thus the numerical values in the new block A43:C45 are identical with those in the loop block A47:C49. To close the loop of iteration we need a cell near the top of the spreadsheet which when copied into A3:C5 will enter in those cells =A47, =B47, ... , = C49. Thus we might write =A46 in cell A2 and to start the iteration copy A2 to A3:C5 (in one copy step).

In *Excel* ™ it is necessary to tell the spreadsheet that you are not self-referencing the spreadsheet by mistake. To do so, go into *Options*, then into *Calculation*, and click on *Iteration*. While in *Calculation* you should also tell the spreadsheet how many iterations you wish to perform and a minimum difference in all cell values in order to continue the iteration (not "maximum," as written incorrectly in the spreadsheet menu). If there is no difference after an iteration or if the difference is less than that specified, then the Excel spreadsheet automatically stops iterating.

All this might be clarified by the Excel spreadsheet for the above damped harmonic oscillator. Initial values and subsequent values appear in the 1x2 rectangular block A3:B3, and Δt appears in C2. The iteration formulas

=A3+(B3)*$C$2 and =B3+(-A3-B3)*$C$2 now appear in A5 and B5, the new block. The loop formulas =A5 and =C5 appear in A7 and B7, the loop block. The formula =A5 appears in A1 so that when A1 is copied into A3:B3 (becoming =A7 and =B7 in those cells), the loop is closed and the iteration begins. Larger models are just extensions of the same process.

	A	B	C
1	1		delta t =
2	x1 =	x2 =	0.1
3	1	0	
4	next x1 =	next x2 =	
5	1	-0.1	
6	loop block =		
7	1	-0.1	

In *Lotus*™ *1-2-3*™ the situation is somewhat different. In *Lotus* the rectangular block of cells with, say, opposite corners A3 and C5 is designated A3.C5 or A3..C5, but the $ convention is the same. Also, it is necessary to write a *Macro* to get a *Lotus* spreadsheet to iterate automatically. A *Lotus* spreadsheet with such a *Macro* is shown on the next two pages, the first page with cell formulas, the second page with numerical values. The system modeled is the same damped harmonic oscillator.

   Start with 0 in the Δt cell,cell B6. Cells A1.A4 constitute the *Macro*. To name the *Macro*, hit *Range, Name, Create*, enter the name \A (note \A, not /A), *Return*, A1.A4, *Return*. To run the model, copy as before to close the loop. For this spreasheet, this means copy from C11 to the variable block A11.B11. Then enter the number for Δt in the designated cell. Then hit *Control* and A simultaneously. The spreadsheet should iterate until the maximum time value in cell B7 is reached. Some computers have different conventions on *Macro* activation and it might be necessary to actually read part of the *Lotus* manual.

```
A1: '{goto}B8~+B6+B8~
A2: '{calc}
A3: '/xi(B8-B7)<0~/xgA2~
A4: '/xq
A6: 'delta t
B6: 0.1
A7: 'max t
B7: 5
A8: 'current t
B8: (F1) +B6+B8
A10: 'x1 =
B10: 'x2 =
A11: +A15
B11: +B15
C11: +C15
```

```
{goto}B8~+B6+B8~
{calc}
/xi(B8-B7)<0~/xgA2~
/xq

delta t         0.100000
max  t          5.000000
current t            5.1

x1  =        x2 =
   -0.048430    0.099425     0.000000
```

References

[CJ] R. Carlson and C. Jeffries, Efficient recognition with high order neural networks, to appear in Proc. SPIE conference 1469 Applications of Artificial Neural Networks, Orlando, 1991.

[CG] M. Cohen and S. Grossberg, Absolute stability of global pattern formation and parallel memory storage by competitive neural networks, IEEE Trans. SMC-13 (1983) 815-826.

[Gi] J. D. Gibson, *Digital and Analog Communications*, Macmillan, New York, 1989.

[GM] C.L. Giles and T.P. Maxwell, Learning, invariance, and generalization in high order neural networks, Applied Optics 26 (1987) 4972-4978.

[G] S. Grossberg, Nonlinear neural networks; principles, mechanisms, and architectures. Neural Networks 1 (1988) 17-61.

[Hi] M. Hirsch, Convergence in neural nets. Proc. First IEEE Int. Conf. on Neural Networks (1987) II 115-125.

[Ho] J. J. Hopfield, Neurons with graded response have collective computational properties like those of two-state neurons, Proc. Nat. Acad. Sci. USA 81 (1984) 3088-3092.

[HSB] M. Hussain, J. Song, J.S. Bedi, Neural network application to error control coding, SPIE Applicatoions of Artificial Neural Networks 1294 (1990) 502-509.

[J] C. Jeffries, Code recognition with neural network dynamical systems, SIAM Review 32 (1990) 636-651.

[JP] C. Jeffries and P. Protzel, High order neural models for error correcting code, Proceedings of the SPIE Aerospace Sensing Conference, Orlando, March 1990.

[JvdD] C. Jeffries and P. van den Driessche, Hypergraph analysis of neural networks, Physica D, 39 (1989) 315-321.

[L] Y. W. Lee, *Statistical Theory of Communication*, Wiley, New York, 1960.

[LD...] Y. Lee, G. Doolen, H. Chen, G. Sun, T. Maxwell, H. Lee, and C.L. Giles, Machine learning using a higher order correlation network, Physica D 22, (1986) 276-306.

[LMP] J. H. Li, A. Michel, and W. Porod, Qualitative analysis and synthesis of a class of neural networks, IEEE Trans Circuits and Systems 35 (1988) 976-985.

[MS] F.J. MacWilliams and N.J.A. Sloane, *The Theory of Error-Correcting Codes*, North-Holland, Amsterdam, 1977.

[M] C. Mead, *Analog VLSI and Neural Systems*, Addison-Wesley, Reading, 1989.

[ML] A. M. Michelson and A. H. Levesque, *Error-Control Techniques for Digital Communication*, Wiley, New York, 1985.

[PT] E. Page and G. Tagliarini, Algorithm development for neural networks, Proc. SPIE Symposium Innovative Sci and Tech, 880 (1988) 11-18. 1989.

[P] V. Pless, *Introduction to the Theory of Error-Correcting Codes*, Wiley, New York, 1982.

[PH] J. C. Platt and J. J. Hopfield, Analog decoding using neural networks, Proc. First Neural Information Processing Systems Conference, Amer. Inst. Physics, Snowbird, Utah (1986) 364-369.

[PPH] D. Psaltis, C.H. Park, and J. Hong, Higher order asssociative memories and their optical implementations, Neural Networks 1 (1988) 149-163.

[R] R. Redheffer, Volterra multipliers II. SIAM J. Alg. Disc. Meth. 6 (1985) 612-623.

[Ru] S. Ruimveld, A high order neural network applied to a pattern recognition problem, master's thesis, Clemson University, April 1990.

[S] D. Sánchez, *Ordinary Differential Equations and Stability Theory: An Introduction*, Freeman, San Francisco, 1968.

[T] G. Tagliarini, Ph.D. dissertation, Clemson University, 1989.

[TH] D. W. Tank and J. J. Hopfield, Simple "neural" optimization networks: and A/D converter, signal decision circuit, and a linear programming circuit, IEEE Trans. Circuits and Systems 33 (1986) 533-541.

[W] J. L. Willems, *Stability Theory of Dynamical Systems*, Nelson, London, 1970.

Index of Key Words
(Key topics are listed in the Table of Contents)

additive neural network models 5
adjacent orthants 62
answer set 17
attractor region 3,12,15,139
attractor trajectory 131,138
autonomous 147
balanced cycle 36
barycenter 34
characteristic polynomial 149
code word 82
coefficient index set 33
coefficient simplex 34
composition function neural network i
connected hypergraph 36
connection matrix 21
constant trajectory 11,130
content addressible memory 1,7,54
cover 23,28
cycle 36
cyclic list of g values 102
decoding 82
derivative of a function 133,134
determinant 148
diagonal matrix entry 148
distance 11,138
dynamical system ii,2,3,
         difference equation 129,133,135
         differential equation 135
dynamical system neural network ii
edge 34
eigenvalues 149
encoding 82
ε-invariant
Euler method 137
factorial 148
finite diameter 139
fixed coefficients 65
forward-difference procedure 137
foundation function 40
Fundamental Theorem of Algebra 149
gain 3,9,53,55
gain function 3,9,55
Hadamard code 83-84
Hamming distance 82
Hamming (7,4) code 82-83
high order neural network 5
Hopfield model 73
Hurwitz test 150
hypergraph 34

identity matrix 148
image product 43,59,103
infeasible 6,16
initial state 6
inside (level sets) 144
layers ii
learning 26
level cylinder 144
level set 144
limit cycle 12,140
linear approximation matrix 13,147
linear dynamical system 146
linearization 146
loop block 154
memory 53,59,103
memory model 7,53,**59**,103,**104**,
memorize 14
modified memory model 70-71
multiproduct 33
nxn matrix (n by n) 148
nested 144
neural network model 9
neural network trajectory 9
neurons 3
new block 154
off (neuron) 16,56,120
off-diagonal matrix entry 148
on (neuron) 16,56,120
orthant relative to T_ε 9,56
outside (level sets) 144
path 35
permutation 148
piecewise linear 9
polynomial 148
quadratic function 5
ramp function 9
rate of change 129,133,134,135
recognition 53
region of attraction 12,139
roots 149
same sign barycenter 34
set selection problem 6,17
sign equivalence class 34
sign of the permutation 148
signal to noise ratio 84,85
stable 11
stability 132,137,139
state space 129
state-time 129
step gain function 23
stored memory 59,103
symmetric vertices 65
system functions 1,4,135
time-dependent difference equations 133
total benefit 123

training ii,1,26,95
trajectory 3,9,135
transition 3,6,62,120
transition zone 9,55
variable block 154
vertex 34
vertices of the n-cube 65
vertes-to-vertex path 35
visible 28
Viterbi code 94,122
Volerra multipliers 36
weights, weight parameters ii,26,36,95

Epilog

Future directions

The theory and application of neural networks could be advanced in the following three ways.

First, because of the small size of ε required in the memory model with large n values, it seems pointless to attempt brute force use of the memory model for all but the smallest code correction and image recognition problems. What is needed is coarse classification followed by finer degrees of classification and final recognition. Put another way, we need a theoretical and practical understanding of how to build a hierarchy of layers memory models with parallel computing in each layer. Convergence in and communication between layers should be pipelined, forming an assembly line for processing, classifying, and recognizing signals.

Second, it seems likely that the memory model could be adapted as a trainable point classifier with considerable power. By using an n-dimensional memory model in (n+1)-dimensional space, it is possible to build a dynamical system which classifies any of the 2^n orthants of the subspace as "yes" (trajectories converging to points with $(n+1)^{st}$ component +1), any of the remaining orthants as "no" (trajectories converging to points with $(n+1)^{st}$ component -1), and any remaining orthants going to nearest neighbor. The "exclusive or" problem of perceptron days is trivial for such a system. Furthermore, it is possible to specify nonorthonormal orthants and other distortions of the regions of attraction. Also, by going to (n+p)-dimensional space, the conclusion of the classifier can be made in the form of any binary vector of length p. Finally, it is possible to train such a system as perceptrons are trained. Such a classifier could become part of an adaptive controller.

Third, there is a need to persue implementation of the memory model and the above generalizations in hardware.

All of the above topics are under current study at Clemson University.

MATHEMATICAL
MODELING

Series editors:

William Lucas
Department of Mathematics
Claremont Graduate School
Claremont, CA 91711

Maynard Thompson
Department of Mathematics
Indiana University
Bloomington, IN 47405

Mathematical Modeling is a series of carefully selected books which present serious applications of mathematics for both the student and professional audience. The series aims to familiarize the user with new models and new methods and to demonstrate the art of constructing useful mathematical models of real-world phenomena.

We encourage preparation of manuscripts in some form of TeX for delivery in camera-ready copy, which leads to rapid publication, or in electronic form for interfacing with laser printers or typesetters.

Proposals should be sent directly to the editors or to: Birkhäuser Boston, 675 Massachusetts Avenue, Cambridge, MA 02139.

MMO 1   *Probability in Social Science,* Samuel Goldberg
MMO 2   *Popularizing Mathematical Methods in China: Some Personal Experiences,* Jua Loo-Keng and Wang Yuan
MMO 3   *Mathematical Modeling in Ecology,* Clark Jeffries
MMO 4   *Newton to Aristotle: Toward a Theory of Models for Living Systems,* John Casti and Anders Karlqvist, eds.
MMO 5   *Introduction to Queueing Theory, 2nd edition,* B. V. Gnedenko and I.N. Kovalenko (translated from Russian)
MMO 6   *Dynamics of Complex Interconnected Biological Systems,* Thomas L. Vincent, Alistair I. Mees, and Leslie S. Jennings, eds.
MMO 7   *Code Recognition and Set Selection with Neural Networks,* Clark Jeffries